This book should be returned to any branch of the
Lancashire County Library on or before the date

Marianne Freiberger and Rachel Thomas are the Editors of *Plus* Magazine (www.plus.maths.org), a free online magazine opening a door to the world of maths for the general public.

Before joining *Plus* in 2005, Marianne did a PhD in pure mathematics, followed by three years as a postdoc at Queen Mary, University of London. She has also been Editor-in-Chief of the mathscareers website.

Rachel worked as a maths consultant for business, government and industry after completing her Masters in pure mathematics at the University of Western Australia. She has edited the Gazette of the Australian Maths Society and designed mathematical walking tours with Marcus du Sautoy for Maths in the City.

Rachel and Marianne were editors of the popular maths book *50: Visions of Mathematics* (OUP, 2014).

NUMERICON

Marianne Freiberger
and Rachel Thomas

Contents

Introduction

Humanity has an urge to explore. We can't stop ourselves from wondering what is around the corner, over the next hill and beyond the horizon. This is as true for the great explorers as it is for any of us travelling somewhere for the first time. In the spirit of exploration, we would like to take you on a journey to an exciting land, one that is familiar to some but foreign to many. We will take you to some of our favourite places, show you the dramatic landscapes, beautiful views and precious treasures, and tell you stories of the valiant heroes, the baffling mysteries and the brilliant conquests. We're going on a guided tour through the world of mathematics.

It's a world that can seem hard to penetrate, but the forbidding symbols and equations are just another language: code for beautiful ideas that often find surprising uses in the ordinary world we all live in. We will help translate, taking you to some of mathematics' most famous landmarks, as well as some secluded coves and exotic beaches we've discovered on our own travels. Our guides will be those friendly representatives of maths we all meet every day: numbers. Each number is an opportunity to stop off, enjoy the view and explore the local territory along paths we've found fun to travel down ourselves.

People love a place for many reasons – the views, the weather, the people, the food, the culture – there are also many things that draw people to mathematics. For many, maths is fundamentally beautiful; indeed many mathematicians won't be entirely satisfied with their work until it has an elegance, simplicity and grace. Others are drawn by its 'unreasonable effectiveness' – its power to explain the world we live in. Sometimes this happens long after a piece of mathematics was discovered and often it's hidden from view. Maths is the language spoken by all the sciences, taking us to the frontiers of knowledge, from the workings of the Universe to the workings of our minds, which enabled us to dream it all up in the first place.

As editors of *Plus* magazine (plus.maths.org), an online magazine that aims to open the door onto the world of maths, we've had the privilege to explore mathematics widely and to meet some of the amazing (and sometimes eccentric) people that build it. Apart from visiting our favourite mathematical sights, we would also like to tell the stories of the people and cultures that have created them. Funny, bizarre, tragic and dramatic, these stories are worth telling all by themselves. And in the same way that you appreciate a great piece of architecture more when you learn who built it and why, these stories can also give great insight into the lofty mathematical structures we'll meet along the way.

This book is our opportunity to do what we love doing most: to show off the beauty of maths in all its glory and tell the stories that weave through it. You'll probably have heard of many of our destinations, but some may be new, and we might even reveal some surprises along the way. We hope you enjoy the ride ...

0 How nothing gave us something

In the beginning, there was nothing. Well, actually, no. In the beginning there was always something. It might have been beans, successful hunts, or victories in battle, but, for millennia, people were using maths to describe things – counting them, measuring them, dividing them up. A mathematical description of nothing, zero, was still a long way off.

Count for something

It's most likely that early humans counted on their fingers in the same way that we all first learn to count. (It's handy having a set of counting sticks at the end of your arm, or, rather, in the fold of your animal skin.) One of the first pieces of evidence of our use of numbers is what are believed to be tally marks cut into a 20,000-year-old bone, known as the *Ishango Bone*, found in Zaire, Africa. A tally system is a very sensible way to keep track of accumulating quantities, whether you're keeping track of a score or you are a prisoner marking your days in prison on your cell wall. The way we keep control of a large number of tally marks today is still firmly connected to our early days of counting – we group them into fives, like the five fingers on our hands.

The first four are marked individually, the fifth as a line crossing the first four, making a complete set. It makes sense that our idea of an easily manageable set is the same as the count of the digits on our hand.

What we called these counting quantities, whether we even had words for them, is another question. There are still cultures today, including the Pirahã and Mundurukú from the Brazilian Amazon, who have a name for small numbers or quantities, but refer to anything larger simply as 'many'.

But over the centuries almost all cultures developed names and symbolic representations for numbers, and a way to combine these to write any number they could possibly need. Inscriptions found in Egyptian tombs from over 5,000 years ago (3000 BC) show that the Egyptians were using beautiful hieroglyphs to represent numbers, such as coils of rope, lotus flowers and frogs, to represent 100, 1,000 and 100,000. These symbols would then be repeated, to build up the number they required, some numbers requiring a large collection of symbols to be represented.

The number 4622 written in Egyptian numerals

The ancient Greeks built up numbers in a similar way, using letters from their alphabet to write numbers – for example α for 1, β for 2, γ for 3, κ for 20, τ for 300. The Romans used combinations of symbols such as I for 1, V for 5, X for 10, L for 50, C for 100, D for 500 and M for 1,000 to write numbers. Generally Roman numbers were built up by adding together the symbols' values, for example XII means 12 (though there were conventions for subtraction too, for example IV means 5 – 1 = 4). It's a system we still use today for names of kings and queens (King George VI) and for dates at the end of movies and TV shows.

But, a number, no matter what language it is written in, is just a name or symbol for the quantity of things that are being counted. The number 3 means the same thing whether it is written in tally marks or in Egyptian, Roman or Greek numerals. The one-ness of a set of one thing, the two-ness of a set of two things, the three-ness of a set of three things is one of the very first mathematical abstractions that we all intuitively make. The number of the things we are counting is independent of what those things actually are, be they kittens or cabbages.

It all adds up

None of the systems we've looked at included a symbol for the set of no things – it just wasn't necessary. All of the systems are *additive* – you simply add up the values of the symbols (or blocks of symbols) in the number to get its value. There may be a convention for ordering in such systems, for example starting with the largest block on the left. But there isn't usually any ambiguity, because reading the number is just about adding up the individual pieces. (For example the Roman number MCMLXXIV can be read as four blocks, M+CM+LXX+IV, meaning 1,000 + 900 + 70 + 4 = 1974.)

This is good, but it does lead to complications if you are dealing with big numbers or trying to do complicated sums.

Take, for example, the numbers MCMLXXIV and XXXIX written in Roman numerals. If you add these together the answer is MMXIII. But this is a far harder task than adding 1974 and 39 (to get 2013) with our modern numerals. This is where tally-based number systems fail. To really get control of large numbers and to simplify mathematical calculations takes a cleverer way of writing numbers. The vital piece to make such a system work is zero.

Babylonian numerals

Place has value

A people we now loosely call the Babylonians lived in Mesopotamia between the Tigris and Euphrates rivers. As far back as 3000 BC the indigenous Mesopotamian people, the Sumerians, used soft clay tablets to write their numbers, imprinted with the wedged end of their writing stylus. This way of writing numbers evolved into a system that used two wedge-shaped symbols arranged to represent all the numbers 1 to 59.

But rather than continue in this way, inventing ever new arrangements and symbols, the Babylonians, some 4,000 years ago, made a brilliant leap: they invented a *place value system* very similar to what we use today. The numbers are written side by side in a string, and the *value* of each number depends on its *place* in this string.

We can illustrate this using our own number system. For us the digit '4' in 4622 no longer represents the value 4. Instead, it tells us that our number contains exactly 4 multiples of 1,000. Similarly, the 6 tells us that there are 6 multiples of 100. And the two 2s in the number represent different values: the left-most 2 means that there are 2 multiples of 10 and the right-most 2 represents 2 multiples of 1. What do the *place values* 1,000, 100, 10 and 1 have in common? They are all *powers* of the number 10, that is, numbers you get by multiplying 10 by itself a number of times:

$$1,000 = 10 \times 10 \times 10 = 10^3$$
$$100 = 10 \times 10 = 10^2$$
$$10 = 10^1$$
$$1 = 10^0 \quad \text{(by mathematical convention)}.$$

The Babylonian system worked in the same way except that, instead of being based on powers of 10, it was based on powers of 60. A digit within a number told you how many multiples of 1, 60, 60² (= 3,600), and so on, there were in the number, based on where in the string the digit appeared.

Something for nothing

The place value system was a great advance. It made it possible to write very large numbers without having to invent new symbols to represent greater and greater orders of magnitude. It also made complicated sums easier: the way we write the numbers does some of the work for us. If one number contained 3 multiples of 60 and another contained 4 multiples of 60, then clearly their sum would contain 7 multiples of 60, telling you exactly what to write in the slot allotted to multiples of 60. The only complication arises when the multiples of 60 in the sum give you something bigger than 60². This is dealt with by carrying digits to the next slot along, as we do in our own system.

But there was a hitch. What would you write when there isn't a multiple of 60, or 60² or some other power of 60 in a given number? For example, a number such as $3,601 = 60^2 + 1$ doesn't have a multiple of 60 in it so it would be missing a digit in the 60 slot. Originally the Babylonians indicated such a missing digit with a space, leaving lots of scope for ambiguity: is this an intended space, or just the result of the writer having a hiccup? The Babylonians seemed able to cope with this ambiguity by an intuitive understanding of the size of the numbers they were dealing with for any particular calculation. But what they really needed was a *placeholder* to separate the powers of 60.

Around 300 BC such a new symbol began to appear in the shape of two angled wedges. Whenever you came across those you knew that a power of 60 was missing. The new sophisticated number system allowed Babylonian mathematics to flourish. Complex calculations now became possible and spawned extremely accurate astronomical tables.

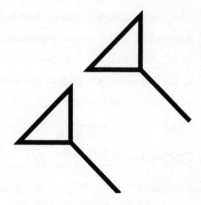

The Babylonian placeholder symbol to separate powers of 60

The place value system was invented at least another couple of times before the origins of our own came along: by the Chinese, from around 300 BC, and by the Mayans, whose culture began as far back as 2000 BC but peaked around AD 500. Both systems also developed a placeholder symbol; zero had begun its inexorable spread in mathematics.

Nothing really is something

What none of these cultures seem to have recognized, however, was that their placeholder symbol – their zero – was a number in its own right. That realization, along with a number system that we use today, comes from India. Indians were using the system, which was positional and had a decimal base, as far back as AD 500. In AD 499 the mathematician and astronomer Aryabhata beautifully captured its essence in his book *Aryabhatiya*:

From place to place, each is ten times the preceding.

Indians used the Sanskrit word for 'void', *s´ūnya*, to refer to zero. Our first record of a little round circle used to describe it is from 870. It later mutated into the symbol for zero we use today.

But, most importantly, Indian mathematicians treated zero as a number in itself, a number they could do calculations with and that might even pop out as the answer to a problem.

In his book *Brahmasphutasiddhanta*, published in around 628, mathematician and astronomer Brahmagupta laid down rules for arithmetic. In doing so he captured the essence of zero's nothingness, at least as far as arithmetic was concerned. We can express it as follows:

When zero is added to a number or subtracted from a number, the number remains unchanged:

$$b + 0 = 0 + b = b, \quad b - 0 = b.$$

This fact makes zero unique among numbers: no other number leaves its partner in addition (or subtraction) quite so undisturbed. For suppose that there *was* another such number and call it *u* for 'unknown'.

Since adding u to any number leaves that number unchanged we have

$$0 = 0 + u.$$

And since adding 0 to any number also leaves that number unchanged, we also have

$$0 + u = u.$$

Putting the two together gives

$$0 = 0 + u = u.$$

So u was equal to 0 all along!

This, incidentally, is our first example of a mathematical proof: an argument that shows, beyond any doubt, that something is true. The concept of proof belongs to mathematics like fish belong to water and we will be meeting it many times later on.

Another rule attributed to Brahmagupta concerns the behaviour of zero under multiplication:

Zero multiplied by any number is zero.

This little number, so unobtrusive when it comes to addition, under multiplication becomes all absorbing.

What, then, can we say about zero and division? What is 5 divided by 0, or 0 divided by 0? These turned out to be tricky questions that planted the seed for mathematics that was developed centuries later. Brahmagupta hedged his bet on the answer to the former, but was categorical about the latter, asserting that 0 divided by 0 (i.e $\frac{0}{0}$) should be 0. In the modern view he was wrong. It was another Indian mathematician who provided a deeper insight into this hairy problem: Bhaskara II.

Fatherly love

Bhaskara II, who lived in the twelfth century AD, is regarded by many as the greatest mathematician and astronomer to emerge from medieval India. His most famous contribution to mathematics, however, seems to have been the result of something we today regard as astronomy's evil twin: astrology. According to legend, Bhaskara consulted his beloved daughter's horoscope and found, to his horror, that she was to remain childless and unmarried. Not prepared to bow to that fate, Bhaskara determined an auspicious moment at which her wedding should take place. To make absolutely sure the moment wouldn't be missed, he constructed a water clock. But his daughter, by the beautiful name of Lilavati, could not suppress her curiosity. When looking at the clock from close up, a pearl from her bridal dress fell into it. It blocked the hole through which the water flowed and thus the auspicious moment could never come. The wedding was off! To console Lilavati, a devastated Bhaskara promised to write a book in her name, a book that would exist forever. Luckily for her, the book was a maths book.

The *Lilavati* is just one part of a greater work, called *Siddhānta Shiromani*, which translates from the Sanskrit as *Crown of Treatises*. It covers an eclectic collection of mathematical questions: there is a lot of arithmetic, but also geometry and algebra. Some questions are directly addressed to Lilavati, 'whose eyes are like fawn's', and many with a poetry our textbooks can only dream of, for example:

> *The square root of half the number of a swarm*
> *of bees is gone to a shrub of jasmine; and so are*
> *eight-ninth of the whole swarm: a female is buzzing to*
> *one remaining male that is humming within a lotus*
> *flower in which he is confined, having been allured*
> *to it by its fragrance at night. Say, lovely woman, the*
> *number of bees.*

If you can't work out the answer, you can find it at the foot of this page.

In the *Lilavati* Bhaskara gives rules for calculating with zero, including one that appears to say that, for any number *a*,

$$\frac{a \times 0}{0} = \frac{0}{0} = a.$$

This seems to suggest that $\frac{0}{0}$ can be anything – any number *a* we care to choose – and we will see an echo of this below. Bhaskara's

---

Answer to Lilavati bee question:
Let x be the number of bees. Then $\sqrt{\frac{x}{2}} + \frac{8}{9}x + 2 = x$, which after some manipulation gives $\frac{1}{81}x^2 - \frac{17}{18}x + 4 = 0$, giving a solution of $x = 72$.

---

great insight, however, came in a lesser-known work of his, the *Vija-Ganita*, where he considers what $\frac{a}{0}$ should be:

> Quotient the fraction $\frac{3}{0}$. This fraction of which the denominator is [zero], is termed an infinite quantity. In this quantity ... there is no alteration, though many be inserted or extracted; as no change takes place in the infinite and immutable God.

So according to Bhaskara, the result of division by zero should be infinity, a number he equates with an unchanging God as infinity is unchanged by addition or subtraction. Modern mathematicians do not agree with this idea, but it is quite easy to see why Bhaskara came up with it. If you divide a line segment into smaller and smaller pieces, the number of pieces gets larger and larger. As the length of your pieces gets closer to zero, the number of them gets closer to, well, infinity.

From above and below

Under normal circumstances division behaves in a nice continuous fashion. If I divide 1 by a sequence of numbers that get closer and closer to 2, then the result will get closer and closer to $\frac{1}{2} = 0.5$:

$$\frac{1}{1.9} = 0.5263\ldots$$

$$\frac{1}{1.99} = 0.5025$$

$$\frac{1}{1.999} = 0.5003$$

and so on. If we assume that the same should happen when dividing a number by numbers that get closer and closer to 0, then we get:

$$\frac{1}{0.001} = 1,000$$

$$\frac{1}{0.0001} = 10,000$$

$$\frac{1}{0.00001} = 100,000$$

$$\frac{1}{0.000001} = 1,000,000$$

This seems to be getting larger and larger, approaching or *tending to* infinity, which suggests that any number divided by 0 gives infinity.

But unfortunately things are not quite as simple as this. Imagine the number 0 as it appears on your thermometer, with positive temperatures above it and negative ones below it. If we mark the sequence of numbers we divided by in the argument above – the smaller and smaller lengths – on the thermometer we would get a sequence of numbers creeping up on 0 from the positive side of the thermometer. But we could equally have crept up on it from the negative side, dividing by negative numbers that get closer and closer to zero, for example –0.001, –0.0001, –0.00001, –0.000001 and so on. Dividing a positive number, such as 1, by a negative one gives you a negative answer, so the results are now

$$\frac{-1}{0.001} = -1,000$$

$$\frac{-1}{0.0001} = -10,000$$

$$\frac{-1}{0.00001} = -100,000$$

$$\frac{-1}{0.000001} = -1,000,000.$$

This sequence seems to tend to something infinite as well, but it seems to be infinite in the opposite direction! Rather than climbing higher and higher on the thermometer we are dropping lower and lower. Is there such a thing as minus infinity? And if there is, is it different from plus infinity? These are difficult questions. Suffice to say that modern mathematicians refuse to commit when it comes to dividing a number by 0: they simply state that the result of such a division is *undefined*.

To the limit

What, then, about dividing 0 by 0? Nothing divided by something is still nothing, that's an uncontentious issue that was already decided by Brahmagupta. So if we divide zero successively by a sequence of numbers that get closer and closer to it we always get 0

$$\frac{0}{0.1} = 0$$

$$\frac{0}{0.001} = 0$$

$$\frac{0}{0.0001} = 0$$

The same is true if we divide by the negatives of the numbers in that sequence:

$$\frac{0}{-0.1} = 0$$

$$\frac{0}{-0.01} = 0$$

$$\frac{0}{-0.001} = 0$$

and so on. So in accord with Brahmagupta, we could be tempted to decide that $\frac{0}{0} = 0$.

But again there is a hitch. What if I take two sequences of numbers, both creeping up on zero, say

$$0.01, 0.001, 0.0001, \ldots$$

and

$$0.02, 0.002, 0.0002, \ldots$$

and divide the corresponding terms by each other? We get

$$\frac{0.02}{0.01} = 2$$

$$\frac{0.002}{0.001} = 2$$

$$\frac{0.0002}{0.0001} = 2$$

$$\frac{0.00002}{0.00001} = 2$$

and so on. Since both sequences, the one that gives us the top of the fractions and the one that gives us the bottom, creep up on 0, this might suggest that $\frac{0}{0}$ should be equal to 2. Equally, if I had turned the division around and divided the numbers in the first sequence by those in the second, the same reasoning would suggest that the result of $\frac{0}{0}$ should be $\frac{1}{2}$! As it turns out, by choosing the two sequences just right you can make a case for any number being the result of $\frac{0}{0}$. Which is why mathematicians have opted out of this one too. The answer of $\frac{0}{0}$ is officially undefined. Nothing divided by nothing is no thing!

Despite these difficulties, and thanks to the initial efforts of Bhaskara and his contemporaries, we are today happy to use zero both as a placeholder in our number system and as a number in itself. And zero has become even more valuable in today's digital age. But to unlock the secret of information we need to combine the power of zero with the number 1.

1 One is all you need

Let's try that again: in the beginning there was 1. And 1 is all you need. Think of a counting number; 1, 2, 3 and so on. You can get to it by repeatedly adding 1s, for example: 2 = 1 + 1, 4 = 1 + 1 + 1 + 1, 7 = 1 + 1 + 1 + 1 + 1 + 1 + 1. It's tedious but it's easy. And if you imagine that for each 1 you add, you take one step along a straight line, then the natural numbers line up neatly and evenly spaced a distance of 1 apart.

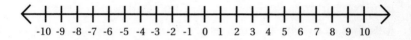

This concept of a number line is incredibly useful. Addition simply becomes moving forward: adding 4 to 6 means starting from 6 and walking 4 steps forward. You would end up at the same place by starting from 4 and walking 6 steps forward, which makes sense as order doesn't matter in addition – it is *commutative*.

Subtraction is moving backwards: 6 – 4 is taking four steps backward from 6. Easy! This also brings negative numbers into the picture. Walking 6 steps backward from 0 gives you –6, and walking 8 steps backwards from 4 gets you to –4. So 4 – 8 = –4. Equivalently you could walk 4 steps forward from –8, showing that 4 – 8 = –8 + 4 = –4. Subtraction, thought of as addition with a negative number, is commutative too.

Even multiplication yields to the power of repeated 1s: 2×4 is taking four steps forward, twice starting from 0; or, equivalently, taking two steps forward, four times (like well-behaved addition, multiplication is also commutative). The multiplication 2×-4 means taking four steps backward, twice, giving you -8, or equivalently, taking two steps backward, four times; which shows that $2 \times -4 = -2 \times 4 = -8$. As you learnt at school, 'positive times positive is positive' and 'positive times negative is negative'.

This leaves us with the one hairy case: what is 'negative times negative'? It's positive, as any textbook will confirm, but why? Teachers often say that's just the way it is. It's a convention adopted to make arithmetic consistent. But the number line gives some intuition as to why this makes sense. A minus sign indicates a reversal of direction: -2×3 means 'take 2 steps backward, three times' but it can also be interpreted as 'take three steps, twice, but take them backward rather than forward'. More generally, $-2 \times$ *some number n* can be interpreted as 'do *n* steps twice, but in the opposite direction to the one indicated by *n*'. So if the number *n* in the bracket is negative, say you are calculating -2×-3, then you are actually taking 6 steps forwards: $-2 \times -3 = 6$.

On or off?

So 1 can get you anywhere as far as the whole numbers and their arithmetic is concerned. But real life is a little more complicated than this. In real life we face choices, of which there are always at least two: left or right, tea or coffee, dog or cat. For machines one particular pair of choices is of special significance: on or off, which we can write as 0 (off) and 1 (on). It turns out that this crucial choice is the only one necessary: the *binary* world is what powers our digital lives.

To see how, start with numbers. As you saw in chapter 0, our way of writing numbers depends on two ingredients: the ten symbols 0 to 9, and their position in a number, which tells us what they mean. The 7 in 7,345 stands for $7 \times 1,000$, the 3 for 3×100, the 2 for 2×10 and the 5 for 5×1. What is so special about 10, 100, 1,000, and so on? They are all powers of 10: $10 = 10^1$, $100 = 10^2$, $1,000 = 10^3$. Even 1 is a power of 10 because by mathematical convention $10^0 = 1$. Our number system is positional and *decimal*, taking the number 10 as its base.

But the choice of 10 is arbitrary. We could just as well have made do with two symbols, 0 and 1, and worked with powers of 2. The number 2, which is $1 \times 2^1 + 0 \times 2^0$, then becomes 10 when written as a *binary number*. The number 3, which is $1 \times 2^1 + 1 \times 2^0$, is 11 in binary. Four, corresponding to $1 \times 2^2 + 0 \times 2^1 + 0 \times 2^0$, is 100 in binary. Here is what all the numbers zero to ten look like written in binary:

0	zero
1	one
10	two
11	three
100	four
101	five
110	six
111	seven
1000	eight
1001	nine
1010	ten

In this way any positive whole number can be represented as a binary string made of 0s and 1s. You can get to the negative numbers by sticking an additional digit in front of such a sequence to act as a minus sign (there are different conventions for doing this). To get to numbers that aren't whole you play the same game, but with powers of $\frac{1}{2}$. For example, the binary expression 0.111 means $1 \times \left(\frac{1}{2}\right)^1 + 1 \times \left(\frac{1}{2}\right)^2 + 1 \times \left(\frac{1}{2}\right)^3 = 0.875$ in decimal. And the binary expression 11.01 means

$$1 \times 2^1 + 1 \times 2^0 + 0 \times \left(\frac{1}{2}\right)^1 + 1 \times \left(\frac{1}{2}\right)^2 = 2 + 1 + \left(\frac{1}{2}\right)^2 = 2 + 1 + \frac{1}{4}$$

which is equivalent to the decimal number 3.25.

Any number that you can write as a decimal can be expressed using only 0s and 1s. And this is indeed the way computers represent numbers.

True or false?

But if computers were all about numbers, they'd be nothing more than glorified calculators. Their real power lies in their ability to perform complex tasks that allow you to book your holiday or survive a boring meeting by illicitly playing Minesweeper. They can do this because they work on one fundamental assumption: that everything is either true or false. In real life this doesn't quite work, but in mathematics it (usually) does. This idea inspired the 19th-century English mathematician George Boole (1815–1864) to build up an entire system of logic.

Boolean logic rests on the idea that statements can be strung together using words like AND and OR. Whether a composite statement is true or false depends on the truth or falsity of its component parts. For example, suppose you know that the statement 'Jim is walking towards me' is true. Then does that make the statement 'Jim is walking towards me AND Jim is dead' true? No, it doesn't. (Unless, of course, Jim is a zombie, in which case, run!) A composite statement P AND Q is only true if both components P and Q are true. Only one of them being true isn't good enough, and both of them being false definitely makes the composite false.

You can capture this in a *truth table* for the AND operator. It looks at all combinations of true and false for the two components P and Q and tells you what the corresponding result for P AND Q is.

P	Q	P AND Q
True	True	True
True	False	False
False	True	False
False	False	False

The OR operator is a lot more permissive. The composite statement P OR Q, say 'Jim is walking towards me OR Jim is dead', is true as long as one of the components is true.

P	Q	P OR Q
True	True	True
True	False	True
False	True	True
False	False	False

Apart from AND and OR, Boole also had a NOT operator, which only takes one statement as input.

P	NOT P
True	False
False	True

If the statement 'Jim is walking towards me' is true then clearly 'Jim is NOT walking towards me' is false and vice versa. So NOT simply switches the truth value of a statement.

Using AND, OR and NOT you can construct all sorts of complicated statements and by doggedly chasing through the truth tables decide whether they are true or false.

Logical sums

All this trailing through truth tables sounds complicated and dreary, but in fact we can simplify things by turning them into a maths problem. Boole ingeniously recognized that binary logical operations behaved in a way that's strikingly similar to our normal arithmetic operations, with a few twists.

First of all, the variables in our new kind of arithmetic (called *Boolean algebra*) are logical statements (loosely speaking, sentences that are either true or false, like 'Jim is a zombie'). As these can only take two values we can write 0 for a statement we know is false and 1 for a statement we know is true. Then we can rewrite OR as a kind of addition using only 0s and 1s:

$0 + 0 = 0$ (since 'P OR Q' is false when both P and Q are false)
$1 + 0 = 0 + 1 = 1$ (since 'true OR false' and 'false OR true' are both true).

$1 + 1 = 1$ (since 'true OR true' is true).

We can rewrite AND as a kind of multiplication:

$0 \times 1 = 1 \times 0 = 0$ (since 'false AND true' and 'true AND false' are both false)
$0 \times 0 = 0$ (since 'false AND false' is false)
$1 \times 1 = 1$ (since 'true AND true' is true).

As the variables can only have the values of 0 and 1, we can define the NOT operation as the *complement*, taking a number to the opposite of its value:

If $A = 1$, then $A' = 0$
If $A = 0$, then $A' = 1$
$A + A' = 1$ (since 'true OR false' is true)
$A \times A' = 0$ (since 'true AND false' is false).

Our new version of these operations is similar in many ways to our more familiar notions of addition and multiplication but there are a few key differences. Parts of equations can conveniently disappear in Boolean algebra, which can be very handy. For example, the variable B in

$$A + A \times B$$

is irrelevant, no matter what value B has or what logical statement it represents. This is because if A is true (or equivalently $A = 1$) then A OR (A AND B) is true no matter whether the statement B is true or false. And if A is false (that is, $A = 0$) then (A AND B) is false no matter the value of B, and so A OR (A AND B) is false. So Boolean algebra provides us with a disappearing act: the expression $A + A \times B$ is equal to a simple little A:

$$A + A \times B = A.$$

Bright sparks

It was the power of simplification that interested Claude Shannon (1916–2001), who in 1936 was a 20-year-old student at Massachusetts Institute of Technology writing his Masters thesis.

Shannon's Masters thesis brought together ideas from his undergraduate degrees in mathematics and electronic engineering. He considered complex circuits of switches and relays that were used in places such as telephone exchanges and realized, thanks to his mathematical training, that these circuits were a physical embodiment of Boole's algebra of logic.

Suppose you have an electrical circuit with a switch and a light bulb attached to it. The light will come on if the switch is on, closing the circuit. When the switch is off the circuit is open and the light is off. Now suppose you have two switches arranged in series (as in the image on the right, 1). The light will come on if both switches are closed. If on the other hand the switches are arranged in parallel (as in the image on the right, 2) then the light is on if either one of the switches is on.

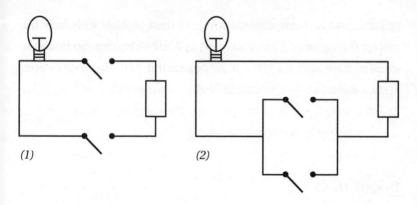

Circuits with switches: (1) switches in series, (2) switches in parallel

We can write all this down in tables:

Circuit 1 – switches in series

Switch 1	Switch 2	Light
On	On	On
On	Off	Off
Off	On	Off
Off	Off	Off

Circuit 2 – switches in parallel

Switch 1	Switch 2	Light
On	On	On
On	Off	On
Off	On	On
Off	Off	Off

These should look familiar. Replace 'On' by 'True' and 'Off' by 'False' in circuit 1 and you get the truth table for the AND operator. Similarly, circuit 2 gives the truth table for the OR operator. You can also construct a circuit for the NOT operator using a switch that acts in reverse to normal switches: when a *normally closed* switch is off it is closed, completing the circuit, and when it is on it is open, breaking the circuit.

This correspondence between circuit and logic is incredibly useful. Suppose that you have a complicated circuit design, a mess of wires and switches. Shannon realized that if you wrote down the corresponding Boolean algebra expression, you could quickly use the simplification laws to remove redundant components in your circuit (see the box for an example).

Before Shannon's work, simplifying a circuit design involved writing all the possible positions of the switches in the circuit and following through the chain of events for each, a process he himself described as 'very tedious and open to errors'. But now, thanks to his insight, any circuit could be easily described and simplified using Boolean algebra.

But Shannon's ideas went even further. In 1948, when he was a research mathematician at the telephone company AT&T Bell, he published the revolutionary paper 'A mathematical theory of communication'. His revolutionary idea was that any information, whether pictures, words, sounds or numbers, could be described mathematically using a series of 0s and 1s. It was here that these binary digits, the 0s and 1s, were described as 'bits' for the first time. Using the electronic circuits that embody the logical AND, OR and NOT operators, you can perform any operation that can be broken down into logical steps on this binary information. You are limited only by your imagination and the technology of the day.

KEEP IT SIMPLE, SHANNON

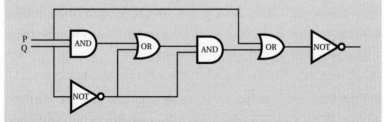

This circuit diagram is written in shorthand with symbols to the AND, OR and NOT gates that combine the 0/1 values for P and Q.

The circuit above corresponds to the expression

$$((P \times Q + Q') \times Q' + P)'.$$

This expression can be simplified using the rules of Boolean algebra to the much simpler expression

$$Q \times P'.$$

So analysing the original circuit using Boolean algebra reveals that a simple circuit with just two gates will do the same trick as the original complicated one.

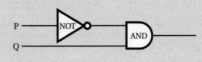

The limits of logic

The on/off world of Boolean logic encapsulates what you might have been thinking about maths all along: things are either true or false and, whichever it is, it has nothing to do with your preference. A few decades after Boole's death, at the beginning of the twentieth century, hopes were riding high that all of maths could be mechanized in the same way as Boole's logic was. The idea was to turn maths in one gigantic but logically rigorous edifice, rather than a loosely connected jumble of different areas such as geometry, algebra, and calculus.

The general idea behind this formal dream goes all the way back to the Greek mathematician Euclid, around 3000 BC. Euclid wrote a book called *The Elements*, which turned out to be one of the most successful maths textbooks of all time; some have claimed that only the Bible produced more editions. In *The Elements* Euclid gave a set of five *axioms* on which he thought all of geometry should be built. These were statements that were so obvious they needed no further justification. His axioms were:

1. A straight-line segment can be drawn joining any two points.
2. Any straight-line segment can be extended indefinitely in a straight line.
3. Given any straight-line segment, a circle can be drawn with the segment as radius and one endpoint as centre.
4. All right angles are equal.

The fifth axiom was a bit of a mouthful, but it is equivalent to

5. The angles in a triangle add up to 180 degrees.

The idea is that any statement you can make in geometry should be derived directly from these axioms using logical arguments. That way, unless you doubt the axioms, the statement is definitely and irrevocably true.

Euclid's axioms were about geometry, but you can also come up with axioms that describe the whole numbers. Without realizing it, we have already come across the central idea behind one such set of axioms, devised by the Italian mathematician Giuseppe Peano (1858–1932) in 1889: that all the natural numbers can be generated by successive steps. Peano's first four axioms were:

- 0 is a number.
- Every natural number has a successor which is also a number.
- No natural number has 0 as its successor.
- Distinct natural numbers have distinct successors.

This gives us all the numbers and their ordering, and also addition and multiplication, which, as described above, correspond to moving backwards and forwards on a number line.

Peano's fifth axiom was not about the numbers themselves but about how you can say something about all of them, even though there are infinitely many. The general idea is that if something is true for 0, the first natural number, *and* you can show that it being true for one number, n, means it is also true for the next one, $n + 1$, then it is true for all of them. It's like a chain of dominoes: if it's true for 0, then it must be true for 1, which means it must be true for 2, and so on, all the way towards infinity. This fifth axiom is called the *principle of induction*.

One set of axioms to bind them all

Peano's work inspired the British thinker Bertrand Russell (1872–1970) to try to complete the axiomatic dream in collaboration with Alfred N. Whitehead (1861–1947). In their monumental work *Principia Mathematica* (published between 1910 and 1913) they intended to show that all of pure mathematics could be built on a small set of concise axioms. This wasn't easy: the proof that $1 + 1 = 2$ doesn't appear until well into the second volume. Neither did they manage to pull it off completely, failing to prove that the axiomatic system they built didn't contain any contradictions.

These exciting developments in logic drew the attention of the Austrian Kurt Gödel (1906–1978), a shy and introverted man with a taste for philosophy. Gödel's work on the subject culminated in a result that came as a shock.

Suppose you have a set of axioms and rules for how to make logical arguments. Also suppose that your axiomatic system is able to express the natural numbers and their arithmetic. And finally, suppose that your system is free from contradiction, which is what you expect from any decent mathematical construct. Gödel's first *incompleteness theorem*, which he proved in 1931, says that within such a system there will always be statements about the natural numbers that you cannot prove to be true or false. By this he didn't mean statements that had nothing to do with numbers, such as 'the world is ruled by shape-shifting lizards' or 'I love you'. He meant statements that can be phrased within the language of the system. There are things that even the best axiomatic system can't prove although it can formulate them. That explains the name of the theorem: any such system is necessarily incomplete.

This does sound bad, but perhaps there is a way to fix it. Suppose you come across an unprovable statement and you know, you are convinced, that this statement should be true. If you can't derive this statement from the axioms, why not simply turn it into one? You simply take it as read and proceed from there. After all, this is what most of us do in real life; there are lots of things we can't prove but choose to believe.

But this quick fix won't work. According to Gödel's theorem, if your new system, with the extra axiom appended, is still free from contradiction, then there will still be other unprovable statements. You simply can't win.

A fundamental flaw?

Think about this for a second and you realize just how devastating the result is. It means that things in maths aren't inherently true or false; they can be neither. In fact, by changing the rules (the axioms), you can make some statements true or false according to personal taste. Perhaps the noble realm of mathematics is simply a matter of opinion?

On the face of it this should get you really worried. After all, the planes you fly in, the car you drive and the tax bill you pay are constructed using mathematics. What if the mathematical truths behind these calculations aren't so true after all? But there is no need to worry. Gödel's original theorem was based around tricky self-referential statements such as 'this sentence can't be proven true or false'. If you could prove it either true or false, then this would make it provable, introducing a contradiction into your system. Since Gödel's time, mathematicians have found more concrete examples of unprovable statements but so far most of these reside in the lofty heights of abstract mathematics without impact on the real world. For the moment, and probably for a few centuries more, real-life mathematics is safe.

Philosophers of mathematics haven't given up either. There may not be a definite set of axioms that hands us all of mathematics on a plate in a strictly true/false fashion. But we can still choose a set that feels most natural; one that renders true those unprovable statements that most chime with intuition. Philosophers too can be practical.

Gödel's incompleteness theorems (there was also a second one) brought him enormous acclaim, but it did not save him from a tragic end. An apparent hypochondriac, he was plagued by a fear of poisoning and, probably imaginary, heart problems. In the mid 1930s, a time in which he travelled between Vienna and the USA, he suffered nervous breakdowns. The person who supported him through these hard times and beyond was his partner, Adele Porkert, a dancer, who was six years older than him, divorced and not too popular with his parents.

In 1939 Gödel was declared fit for service in the German armed forces, a fact that woke him up to the grim reality of Nazi Germany. With help from the Institute for Advanced Study in Princeton, he and Adele were able to obtain visas for travel to the USA in 1940, and they stayed in that country for the rest of their lives. Albert Einstein was a considerable support to Gödel during the late 1940s and early 1950s.

From about 1958, the year in which he published his last paper, Gödel became increasingly withdrawn and mentally unstable. In the mid 1970s he was dealt a couple of cruel blows. In 1976 Adele suffered a stroke which left her in need of his care and around the same time his good friend Oskar Morgenstern died of cancer. Still paranoid of being poisoned, Gödel began to starve himself and eventually died on 14 January 1978.

Gödel's fate is not the only tragic one in mathematics; you will meet quite a few more in this book. Perhaps the most famous one goes back to the ancient Greeks, and it involves a very special number: the square root of 2.

$\sqrt{2}$ Butterflies, murder and a proof that didn't fit in the margin

I f there is one mathematical result you remember from school it's probably Pythagoras' theorem. It goes like this: take a right-angled triangle and construct a square on each of the two sides of the triangle that enclose the right angle. Then the sum of the areas of these two squares is equal to the area of the square constructed on the remaining side. The area of a square whose sides have length a is $a \times a = a^2$. If our triangle has sides of length a, b and c, with c being the longest, then Pythagoras' theorem tells us that:

$$a^2 + b^2 = c^2.$$

This beautiful result lets you compute all sorts of things, including, for example, the length of the diagonal of a square. Together with two sides of the square the diagonal forms a right-angled triangle. If the sides of the square have length 1, then Pythagoras' theorem tells us that

$$1^2 + 1^2 = 2 = d^2.$$

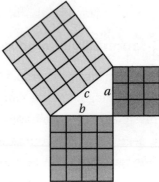

The sum of the areas of the two smaller squares is equal to the area of the larger square.

This means that d, the length of the diagonal, is equal to $\sqrt{2}$ – the number which when multiplied by itself gives 2 as the result.

Slippery root

There is nothing particularly alarming about this until you realize that $\sqrt{2}$ is rather difficult to pinpoint. If you multiply 1.5 by itself you get 2.25, which is too big. Try 1.4 and you get 1.96, which is too small. $(1.41)^2 = 1.9881$ is too small again while $(1.42)^2 = 2.0164$ is too big.

It seems like you can't win and indeed you can't. $\sqrt{2}$ is an *irrational number*, which means that you can never write down all its digits: its full decimal expansion is infinitely long and it doesn't end in a repeating pattern of digits. The first 20 digits are:

1.41421356237309504488

Irrationality can be deadly

This is an intriguing property for a number that arises so innocently as the diagonal of a simple square. And indeed, Pythagoras' disciples were not happy. The Pythagorean brotherhood was a secret cult that operated in the fifth century BC in Croton (modern-day Italy) and, besides vegetarianism and not eating beans, it championed intellectual pursuits as the basis of a morally sound life. Mathematics was at the core of the Pythagorean philosophy: the very words mathematics ('that which is learnt') and philosophy ('love of wisdom') are said to have been coined by Pythagoras, and his motto, according to legend, was 'all is number'.

The trouble was that by 'number' the Pythagoreans meant only the whole numbers and their *ratios*, the fractions $\frac{1}{2}$, $\frac{1}{4}$, $\frac{3}{4}$, etc. Irrational numbers cannot be written as fractions; in fact, this is what defines them (if you know long division you can check for yourself that any fraction expressed as a decimal number yields a finite string of digits, or at least one that ends in a repeating pattern). The discovery that numbers, such as $\sqrt{2}$, can be irrational, attributed to Hippasus of Metapontum, himself a Pythagorean, amounted to nothing short of heresy. According to (admittedly rather murky) historical evidence, his punishment was severe. Hippasus was condemned to death at sea, sunk in a shipwreck. How many other people have died for the sake of a number?

Irrational but not absurd

The standard proof that $\sqrt{2}$ is an irrational number is the prime example of a form of argument that is often used in mathematics: *reductio ad absurdum* or *proof by contradiction* (see the box opposite for the full proof). To prove something (say that $\sqrt{2}$ is irrational) you assume that the opposite is true (that $\sqrt{2}$ can be written as a fraction). If this then leads you to a contradiction, you conclude that your assumption must have been false, so your initial statement, the one you wanted to prove, must be true. This is a natural way of reasoning: if the assumption that the butler committed the murder leads you to conclude that he must have been in two places at once, which is nonsense, then you deduce that the assumption must have been false: the butler is innocent. Proofs by contradiction are a staple in mathematics but they can also have surprising consequences. You will find out more in Chapter 3.

Hippasus' discovery was only the tip of an enormous iceberg. There are infinitely many irrational numbers in any piece of the number line, no matter how small. And while the rational numbers, those that can be written as fractions, can be lined up in an orderly queue and given the labels 1, 2, 3, and so on, the irrationals are so numerous you just can't separate them out like that. If you were to randomly stab a sharp pencil at the number line the probability of hitting an irrational number would be 1 and the probability of hitting a rational one would be 0. So as far as numbers were concerned, the Pythagoreans couldn't have been more wrong.

PROOF THAT ROOT 2 IS IRRATIONAL

Assume that $\sqrt{2} = \frac{m}{n}$ with whole numbers m and n having no common divisors (no number except 1 divides both m and n). Then

$2 = \frac{m^2}{n^2}$,

so $2n^2 = m^2$.

This means that m^2 is even, which in turn means that m is even as the square of an odd number is always odd. So we can write m as $2k$ for some positive whole number k. Substituting $2k$ for m in the equation above gives

$2n^2 = m^2 = 4k^2$.

Dividing by 2 gives

$n^2 = 2k^2$,

so n^2 is even too. But this means that n itself is also even. This is a contradiction because we assumed that m and n don't have any common factors. Therefore $\sqrt{2}$ cannot be written as $\frac{m}{n}$ and is an irrational number.

$\sqrt{2}$ can be good

But the Pythagoreans might not have been so upset at the discovery of irrational numbers had they realized how useful they can be. You come across $\sqrt{2}$ almost every day when you are dealing with sheets of paper. The standard paper sizes used in Europe, A5, A4, A3 and so on, have the very nice feature that two sheets of the same size placed next to each other, say two A4 sheets, give you the sheet the next size up, in this case A3. The sheets fit together so that twice the width W of the smaller sheet gives you the length of the larger one, while the length L of the smaller sheet becomes the width of the larger one.

The aspect ratio of a sheet, that's the ratio between the lengths of its two sides, is the same for all sizes, giving

$$\frac{W}{L} = \frac{L}{(2W)}.$$

This can be rewritten as

$$\left(\frac{L}{W}\right)^2 = 2,$$

which means that

$$\frac{L}{W} = \sqrt{2},$$

The fact that $\sqrt{2}$ is the aspect ratio of every sheet is the defining feature of the A series.

Why is this useful? If you want a photocopier to be able to reduce a sheet of paper to the next size down (or blow it up to the next larger size) you need the sizes in the series to have the same aspect ratio – if they don't then there will be blank margins around the reduced version of your sheet. The fact that in the A series two sheets of one size placed next to each other give you the next size up means that the same reduction rate can be used whether you're trying to reduce two sheets of A4 or one sheet of A3 to the next size down.

Your photocopier knows all this. When it comes to reducing the size of what you are copying, it will offer you a reduction of 70%, or sometimes it's 71%. Writing these in decimal form (as the result of 70 or 71 divided by 100) gives 0.7 and 0.71, both of which are pretty close to

$$\frac{1}{\sqrt{2}} = 0.707106\ldots$$

This reduction rate is just what you need to reduce one A3 sheet (or two A4 ones) to one A4 sheet: the length L and the width W of the original are reduced to $\frac{L}{\sqrt{2}}$ and $\frac{W}{\sqrt{2}}$, which means that the area of the new sheet is

$$\frac{L}{\sqrt{2}} \times \frac{W}{\sqrt{2}} = \frac{LW}{2}.$$

That's half the area of the original sheet, and since the aspect ratio remains the same, we have reduced the original to exactly A4 size.

The same works when you are blowing sheets up. The photocopier will offer you a magnification of 140%, or 141%, which corresponds to a number close to $\sqrt{2}$, so you can blow an A4 sheet up to A3 size.

Snowballs . . .

Photocopiers only approximate the value of $\sqrt{2}$ since they cannot store its infinitely many digits. Surely such approximations should be good enough for any real-life use of numbers? After all, taking just the first five decimal places gives us an accuracy of 1 part in a 100,000.

Not quite. In 1961 careless treatment of decimal places led the meteorologist Edward Lorenz to accidentally create a new branch of mathematics. It's a story that is beautifully described in James Gleick's book *Chaos: Making a New Science*. Lorenz was using a computer to simulate the weather. He had come up with a set of equations that described the bare bones of atmospheric processes – relating quantities such as pressure, temperature and wind speed – and his computer was chugging through the calculations. Given a set of initial values to start off with, it would churn out the numbers that described the weather a day later, then two days later, then three days later, and so on. Lorenz's model was far too simple to describe real-life weather accurately – it was only a toy model – but that is the path of science: you start with what you can get a handle on, compare your results with reality and then tinker, fiddle and refine, or come up with a new idea.

Lorenz's simulated weather, however, looked encouragingly realistic. One day he decided to run a particular simulation, based on a particular set of starting values, once again. But rather than running it from the beginning he took the numbers the computer had produced mid-way through the first run, describing the weather at a particular time some way into the simulation. Feeding those in as starting values for the second simulation should not make a difference to the subsequent evolution of the toy weather: the same calculations should after all spit out the same numbers. But to his surprise Lorenz saw a wildly different weather pattern emerge. Upon checking he realized the problem was down to a tiny inaccuracy in the numbers he had fed in for the second run: rather than 0.506127, which is what the computer had been working with during the first run, he had used 0.506. In the many repeated calculations this tiny error had snowballed to dramatic effect.

. . . and butterflies

The phenomenon that Lorenz discovered is now known as the *butterfly effect* – even the tiny disturbance caused by the flap of a butterfly's wing can grow into a tornado halfway around the world. Lorenz's system of equations suffered from *sensitive dependence on initial conditions* which meant that slightly different starting conditions could yield a vastly different outcome.

Lorenz's discovery, which helped spawn the area of maths loosely called *chaos theory*, is what hampers prediction of all sorts of processes, from the climate to the stock market. Since you are never going to know initial conditions with 100% accuracy, you can't be sure your error isn't going to grow huge. As physicist Niels Bohr is reputed to have said, 'prediction is very difficult, especially about the future'.

Margin for error

Returning to Pythagoras' theorem, not all numbers that arise from it are irrational. Consider, for example, a right-angled triangle in which the sides enclosing the right angle have lengths 3 and 4. Pythagoras' theorem now tells us that the third side, d, satisfies

$$d^2 = 3^2 + 4^2 = 9 + 16 = 25,$$

so $d = \sqrt{25}$. Happily the square root of 25 is not irrational, in fact it's a whole number: 5. The numbers (3, 4, 5) form what is known as a *Pythagorean triple*: a triple of whole numbers (a, b, c) that satisfy the equation

$$a^2 + b^2 = c^2.$$

There are many more such Pythagorean triples, in fact there are infinitely many; (5, 12, 13), (8, 15, 17) and (7, 24, 25) are examples.

Now what if we vary the problem slightly and look for triples of whole numbers satisfying

$$a^3 + b^3 = c^3,$$

or

$$a^4 + b^4 = c^4,$$

or

$$a^5 + b^5 = c^5,$$

or more generally,

$$a^n + b^n = c^n$$

for some positive whole number n?

It's a kind of problem that would naturally occur to anyone who likes playing with numbers and it did occur to the Frenchman Pierre de Fermat (1601–1665). Fermat had become convinced that such triples don't exist when the power in the expression is greater than 2 and he noted this fact in the margin of a maths book:

> *It is impossible to separate a cube into two cubes, or a fourth power into two fourth powers, or in general, any power higher than the second, into two like powers. I have discovered a truly marvellous proof of this, which this margin is too narrow to contain.*

Fermat did come up with an argument that could sort out the case $n = 4$, but that was it. His tantalizing conjecture, which became known as *Fermat's last theorem,* was to haunt mathematicians for over 350 years.

Wily labours

One of the haunted was a ten-year-old Andrew Wiles who discovered Fermat's last theorem in a library book in 1963. He was hooked immediately and later decided to turn the problem into his life's work. After earning a degree and PhD in mathematics from Oxford and Cambridge he eventually established his career with a professorship at Princeton University, New Jersey. Here he started labouring away at the problem in secret; an admission that he was working on this super problem, this holy grail of mathematics, would have drawn too much attention. 'You can't really focus yourself for years unless you have this kind of undivided concentration which too many spectators will have destroyed,' he told the BBC's *Horizon* programme in 1996.

Wiles exploited results that go way beyond a simple shuffling around of whole numbers. If Fermat's equation $a^n + b^n = c^n$ does have whole number solutions *a, b* and *c* for some whole number *n* then it can be related to a geometric object, called an *elliptic curve*. In the 1950s the Japanese mathematicians Goro Shimura and Yutaka Taniyama had suggested that elliptic curves could be related to other kinds of objects called *modular forms*, which are characterized by the symmetries they possess. But it turns out that if Fermat's last theorem were false, then its elliptic curve would *not* match to any modular form.

You can turn this reasoning on its head: if Shimura and Taniyama's conjecture is true, and all elliptic curves are indeed linked to a modular form, then Fermat's last theorem has no option but to be true too. Thus, a proof of the Shimura–Taniyama conjecture would amount to a proof of Fermat's last theorem.

The secret solution

It was this approach that Andrew Wiles chose to take during his period of mathematical seclusion. He set out to prove a special case of the Shimura–Taniyama conjecture that would be sufficient for Fermat's last theorem, undeterred by the prevailing opinion that the conjecture was essentially unprovable using current mathematical expertise. On 23 June 1993, after seven years of work, Wiles reported success. He announced a proof to a riveted audience of mathematicians at the Isaac Newton Institute in

Cambridge, who were well aware of the lecture's monumental importance – you don't see a centuries-old maths problem solved in front of your eyes every day. Alas it turned out that Wiles' proof contained a serious flaw. It took another year of effort and help from one of Wiles' former students, Richard Taylor, to fix it. But in the end the 357-year-old problem submitted to Wiles' techniques.

We will never know what Fermat had in mind when he scribbled into his famous margin. And we can only guess at what Pythagoras would have thought had he known how much was to flow from his beautiful insight into right-angled triangles. Perhaps the beauty and fascination of the two millennia of mathematics that was to follow would have alleviated the Pythagoreans' horror at Hippasus' proof. And as it turns out $\sqrt{2}$ isn't the worst in terms of irrationality. That honour goes to another number: ϕ.

ϕ From irrationality to the divine

As bad discoveries go, Hippasus' was almost as bad as it gets. Not only is $\sqrt{2}$ an irrational number, it is also more irrational than others. But how irrational can an irrational number be? And what is the most irrational number?

An irrational number can't be written as a simple fraction $\frac{a}{b}$ of whole numbers a and b. But we can approximate an irrational number with a simple fraction, for example, $\frac{17}{12} = 1.416666\ldots$ is a pretty close approximation to $\sqrt{2} = 1.414213\ldots$ and $\frac{41}{29} = 1.41379\ldots$ is an even closer one. How do we find such approximations?

The secret lies in an alternative way of writing any number. Let's start with our first *rational* approximation of $\sqrt{2}$: $\frac{17}{12}$. You can write this fraction as

$$\frac{17}{12}$$

$$= 1 + \frac{5}{12}$$

$$= 1 + \frac{1}{\left(\frac{12}{5}\right)}$$

$$= 1 + \frac{1}{2 + \frac{2}{5}}$$

$$= 1 + \frac{1}{2 + \frac{1}{\left(\frac{5}{2}\right)}}$$

$$= 1 + \cfrac{1}{2 + \cfrac{1}{2 + \frac{1}{2}}}.$$

This series of nested fractions is called a *continued fraction*. You can write $\frac{41}{29}$ in a similar way:

$$\frac{41}{29}$$

$$= 1 + \frac{12}{29}$$

$$= 1 + \cfrac{1}{\left(\frac{29}{12}\right)}$$

$$= 1 + \cfrac{1}{2 + \frac{5}{12}}$$

$$= \ldots$$

$$= 1 + \cfrac{1}{2 + \cfrac{1}{2 + \frac{1}{2 + \frac{1}{2}}}}.$$

You can see that the continued fraction for $\frac{41}{29}$ has one more layer to its nest of fractions than $\frac{17}{12}$. You can also see a pattern emerging, which you can continue by including more layers. In fact you can make the result get as close to $\sqrt{2}$ as you like. The continued fraction expansion of $\sqrt{2}$ itself is:

$$\sqrt{2} = 1 + \cfrac{1}{2 + \cfrac{1}{2 + \cfrac{1}{2 + \cfrac{1}{2 + \cfrac{1}{2 + \frac{1}{2 + \cdots}}}}}}.$$

Unlike its rational approximations, this expansion for $\sqrt{2}$ has an infinite number of layers; it never ends. This is true of all irrational numbers – their continued fractions are infinite. By chopping off the infinite expansion of an irrational number at successive layers you get a sequence of fractions approximating it more and more closely.

Continued fraction approximation		
$1 + \dfrac{1}{2}$	$\dfrac{3}{2}$	1.5
$1 + \dfrac{1}{2 + \frac{1}{2}}$	$\dfrac{7}{5}$	1.4
$1 + \dfrac{1}{2 + \frac{1}{2 + \frac{1}{2}}}$	$\dfrac{17}{12}$	1.41666 . . .
$1 + \dfrac{1}{2 + \frac{1}{2 + \frac{1}{2 + \frac{1}{2}}}}$	$\dfrac{41}{29}$	1.41379 . . .
$\vdots$		$\vdots$
$1 + \dfrac{1}{2 + \frac{1}{2 + \frac{1}{2 + \frac{1}{2 + \frac{1}{2 + \ldots}}}}}$	$\sqrt{2}$	1.41423 . . .

The continued fraction for $\sqrt{2}$ reveals a beautiful pattern that is lost when $\sqrt{2}$ is written as a decimal number. But the elegant simplicity of this pattern is also what makes $\sqrt{2}$ awkward in terms of approximation. The continued fraction of an irrational number has a lot to do with how well the number can be approximated by fractions. If the numbers that appear in the continued fraction never get bigger than some bound, then there's an inherent limit to how close each fraction in the continued fraction approximation can get to the irrational number. That's true for $\sqrt{2}$ because the numbers appearing in its continued fraction don't get bigger than 2: $\sqrt{2}$ is *badly approximable*.

So what if we make things worse and consider the number that only has only 1s in its continued fraction:

$$\phi = 1 + \cfrac{1}{1 + \cfrac{1}{1 + \cfrac{1}{1 + \cfrac{1}{1 + \cfrac{1}{1 + \cfrac{1}{1 + \cdots}}}}}}$$

This number, which has the uninspiring decimal value of 1.6180339..., is the legendary *golden ratio*, ϕ. And as you might have guessed, ϕ is the worst approximable irrational of them all – in this sense it is the most irrational of all irrationals.

Blame the rabbits

Out of all the sequences of fractions you could choose to approximate ϕ, the one you get by chopping off its continued fraction can be considered the best. And it also reveals a fascinating and possibly familiar pattern.

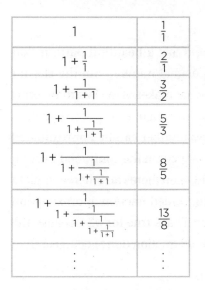

1	$\frac{1}{1}$
$1 + \frac{1}{1}$	$\frac{2}{1}$
$1 + \cfrac{1}{1 + 1}$	$\frac{3}{2}$
$1 + \cfrac{1}{1 + \cfrac{1}{1 + 1}}$	$\frac{5}{3}$
$1 + \cfrac{1}{1 + \cfrac{1}{1 + \cfrac{1}{1 + 1}}}$	$\frac{8}{5}$
$1 + \cfrac{1}{1 + \cfrac{1}{1 + \cfrac{1}{1 + \frac{1}{1 + 1}}}}$	$\frac{13}{8}$
$\vdots$	$\vdots$

The top of one fraction (called the *numerator*) becomes the bottom of the next (the *denominator*):

$$\frac{1}{1}, \frac{2}{1}, \frac{3}{2}, \frac{5}{3}, \frac{8}{5}, \frac{13}{8}, \frac{21}{13}, \frac{34}{21}, \cdots$$

You might recognize the sequence of numbers that wind their way through these approximations of ϕ:

$$1, 1, 2, 3, 5, 8, 13, 21, 34, 55, 89 \ldots$$

They are better known as the *Fibonacci numbers*.

This sequence, one of the most famous sequences in maths, is named after the man who introduced it to the West. Well, not quite. Fibonacci was never a name he used himself. Instead he called himself Leonardo Pisano, after his home town of Pisa in Italy. We think the name that we now know him by, Fibonacci, came from his famous book *Liber Abaci*. The first line of the book roughly translates as 'Here begins the Book of Calculation composed by Leonardo, son of the Bonaccios, Pisano in the year 1202', referencing his father Guglielmo Bonaccio's family. Centuries later, when scholars were analysing hand-written copies of *Liber Abaci* (the only way to publish books in the 13th century was by writing them out by hand as the printing press was still over two centuries away), they seem to have shortened the Latin phrase for this family connection, 'filius Bonacci', to Fibonacci, and taken it as his surname.

It seems somehow appropriate that Fibonacci is now known by a name not his own, given that what he is famous for is a mere example used to illustrate his far greater mathematical achievement. Having grown up and been educated in North Africa, where his father was doing business, and having travelled to Egypt, Syria and Greece, he realized the number system first developed in India and then adopted by the Arabs – using the digits 0, 1, 2, . . ., 9 – was far more useful than the cumbersome Roman numerals still used in the West. When he returned to Italy he wrote *Liber Abaci*, meaning 'the book of calculation', which helped to introduce this number system to the West and demonstrate its worth (which we saw in Chapter 0). What became known as the Fibonacci sequence was one of his examples of calculating with these new numerals:

A certain man had one pair of rabbits together in a certain enclosed place, and one wishes to know how many are created from the pair in one year when it is the nature of them in a single month to bear another pair, and in the second month those born to bear also.

Let's go through this slowly. We start with one pair of baby rabbits. The next month we still have one pair, but as they are now adults they mate, so the third month we have two pairs: one adult pair and one baby pair. The next month the new baby pair is fully grown, so we have two adult pairs, and the adult pair from last month give birth to another pair of babies, so we have three pairs of rabbits all together. Continuing in this totally unrealistic way – babies taking one month to mature, and from then on producing a new pair of babies every month – we get Fibonacci's famous sequence:

	Baby pairs	Adult pairs	Total pairs
Month 1	1	0	1
Month 2	0	1	1
Month 3	1	1	2
Month 4	1	2	3
Month 5	2	3	5
Month 6	3	5	8
Month 7	5	8	13
Month 8	8	13	21
Month 9	13	21	34
Month 10	21	34	55
Month 11	34	55	89
Month 12	55	89	144

At the end of the twelfth month, Fibonacci's highly unrealistic rabbits have produced 144 pairs from a single pair. And intriguingly the Fibonacci numbers appear in all three columns – the number of baby pairs, the number of adult pairs and the total number of pairs of rabbits – each trailing the other by one month.

This intriguing fact is due to the recursive nature of the sequence. The total number of adult pairs in one month is the total number of pairs (adults and babies) in the previous month. The total number of baby pairs in one month is the number of adult pairs in the previous month. So the total number of pairs in one month is

total number of pairs this month
= number of adult pairs + number of baby pairs
= total number of pairs one month ago + number of adult pairs one month ago
= total number of pairs one month ago + total number of pairs two months ago

So each number in the sequence is the sum of the previous two numbers. The Fibonacci sequence was the first such recursive sequence to be studied in Europe.

The Fibonacci sequence seems to capture some essence of growth. Compared with the hypothetical rabbits who revealed the sequence to Fibonacci, a more realistic example of the Fibonacci sequence in population growth is in honey bees. In a hive there is one female queen bee who lays all the eggs. Unfertilized eggs produce male bees, called *drones*. Eggs fertilized by a male bee produce female bees, called *worker bees*, who do not lay any eggs themselves unless they are fed a substance called royal jelly that turns them into egg-laying queen bees to replace an old queen or start a new hive. So, looking at the family tree of a male bee, they have 1 parent, 2 grandparents, 3 great-grandparents, 5 great-great-grandparents, 8 great-great-great-grandparents, and so on. Similarly, a female bee has 2 parents, 3 grandparents, 5 great-grandparents, 8 great-great-grandparents and so on, the Fibonacci sequence counting out the number of ancestors in each preceding generation for any bee.

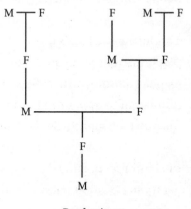

Bee heritage

It's good to be really irrational

One of the best-loved and often repeated facts about the Fibonacci numbers is that you can find them in sunflowers. In fact, you can find Fibonacci numbers in the seed heads and petals of many flowers. Flower seed heads usually contain tightly packed spirals of seeds winding both clockwise and anticlockwise. If you count around the spirals starting at the outside of a seed head (or around any circle within the seed head), the number of clockwise and anticlockwise spirals will almost always be a pair of consecutive Fibonacci numbers. This apparently mysterious appearance of Fibonacci numbers in the seed heads of many flowers, and similarly in pinecones, pineapples and flower petals, is down to their connection with the golden ratio ϕ.

When plants produce new growth – such as branches, leaves or seeds – they do it in a very clever way. For example, when seeds are produced in a seed head they are produced by the growth tip at the centre, called the *meristem*. The seeds are regularly spaced around the meristem, newer seeds pushing older seeds to the outside. If the spacing between one seed and the next is a *rational number*, a simple fraction such as $\frac{1}{2}$, $\frac{1}{3}$ or $\frac{3}{5}$, the seeds gradually line up: if they are half of a turn apart, then after just two seeds they start lining up, each seed being produced exactly in line with a previous seed. A rational turn produces lines of seeds radiating out from the centre of the seed head, leaving large spaces in between.

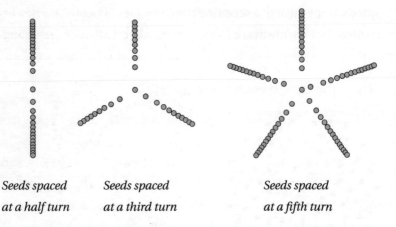

Seeds spaced at a half turn

Seeds spaced at a third turn

Seeds spaced at a fifth turn

This is clearly an inefficient way to pack seeds into the seed head. If the turn between seeds is irrational, the results are better because no amount of turning will result in two seeds being produced in the same position around the centre. The 'more irrational' the number, the better the packing, so the most efficient packing of seeds is produced by φ.

Many plants, most famously sunflowers, have already figured this out and use a turn determined by ϕ. Each new seed is 0.618 (this is ϕ-1) of a turn away from the previous one. And as consecutive pairs of Fibonacci numbers are involved in the closest approximations of ϕ, these are usually the numbers of spirals at the edges of the seed head.

Similarly for many plants producing leaves around the stem of a plant, there are ϕ turns between leaves. This arrangement of leaves allows the most leaves to be exposed to sunlight and rain. Flower petals are really modified leaves, so they also often follow the same spacing. And in the same way as for the number of spirals at the edge of a seed head, the number of petals in a flower is often (but not always) a Fibonacci number. The same reasoning

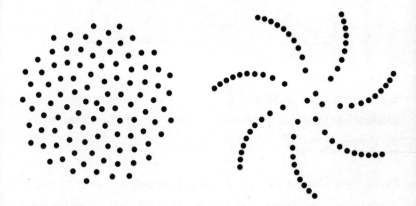

The packing of seeds produced by ϕ turns between seeds (on the left) is better than that produced by π turns between seeds (on the right), since ϕ is badly approximable and π isn't. (Produced with a Wolfram Demonstration)

makes φ an important number in the design of radio telescopes, which need to optimize the way they sample their surroundings for radio signals, just as plants need to optimize their exposure to sunlight and rain.

Mathematical truths and beauty

The special nature of φ had been recognized for millennia; as far as we know it made its first written appearance in Euclid's *Elements*. Euclid referred to φ as an *extreme and mean ratio* and he defined it in the following way:

> *A straight line is said to have been cut in extreme and mean ratio when, as the whole line is to the greater segment, so is the greater to the less.*

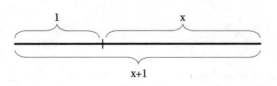

If we take the shorter segment as being length 1 and the longer section to be of length x, then by Euclid's definition we have:

$$\frac{x}{1} = \frac{x+1}{x}.$$

Multiplying both sides of the equation by x keeps the two sides equal, so we get

$$x^2 = x + 1.$$

Thanks to the quadratic formula we learnt at school (see Chapter 12), we know that this equation has two solutions, one of which is

$$x = \frac{(1 + \sqrt{5})}{2}.$$

And this, it turns out, is exactly equal to ϕ. The other solution is $\frac{1 - \sqrt{5}}{2}$ which happens to be equal to $1 - \phi$.

The equation

$$\phi^2 = \phi + 1,$$

shows that ϕ has an unusual mathematical property: it seems to be the place where addition and multiplication meet. Multiplying ϕ by itself is the same as adding 1. If you divide both sides of the equation by ϕ you get

$$\phi = 1 + \frac{1}{\phi},$$

which you can rewrite as

$$\frac{1}{\phi} = \phi - 1.$$

So taking ϕ's *multiplicative inverse*, $\frac{1}{\phi}$, is the same as subtracting 1 $\left(\frac{1}{\phi} = \phi - 1\right)$! You can also spot this directly from the continued fraction expansion of ϕ:

$$\phi = 1 + \cfrac{1}{1 + \cfrac{1}{1 + \cfrac{1}{1 + \cfrac{1}{1 + \cfrac{1}{1 + \cfrac{1}{1 + \cdots}}}}}}$$

so

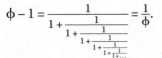

$$\phi - 1 = \cfrac{1}{1 + \cfrac{1}{1 + \cfrac{1}{1 + \cfrac{1}{1 + \cfrac{1}{1 + \ldots}}}}} = \frac{1}{\phi}.$$

Most of ϕ's occurrences in nature can be traced back to mathematical properties such as these.

So fascinated were mathematicians with these curious properties that Luca Pacioli (1445–1519) declared ϕ to be the 'divine ratio' in his book by the same name, which was illustrated by none other than Leonardo da Vinci. Many then and now believe that ϕ holds the secret to beauty and can be found repeatedly in art and architecture, with lengths that appear together, for example as sides of a rectangle, having the ratio ϕ. But, while it is true that psychological studies have claimed that humans find this ratio the most aesthetically pleasing, there is actually little evidence that artists (including da Vinci) explicitly used the ratio in their work.

The golden spiral

There is one more mathematical curiosity about ϕ that is worth looking at. To see it, start by drawing two squares of side length 1, side by side. Above the two you can now neatly fit a square of side length 2. To the left, fit a square of side length 3, then below a square of side length 5, and so on. This is called a *Fibonacci tiling* because the side lengths are successive numbers in the Fibonacci sequence.

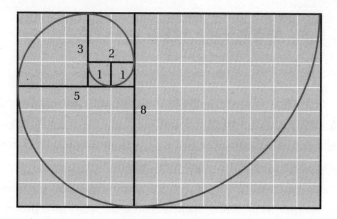

Fibonacci tiling

Now inscribe a quarter circle into each of the squares that passes through two diagonally opposite corners, making sure that circles through squares that correspond to successive Fibonacci numbers line up. What you get is a spiral that looks quite a lot like spirals we observe in nature, for example in nautilus shells, which grow compartment by compartment. It turns out that this spiral approximates something called the *golden spiral*, which has a winding growth that is intimately related to ϕ. You will find out more in our chapter on another fundamental constant of mathematics: *e*. But first we will meet what is, in some sense, the first really interesting whole number.

2 Of prime importance

Among the natural numbers, 2 is the first to give us more than the bare minimum. As we saw in Chapter 1, we could imagine the natural numbers the boring way, as beads on a string: a bead for 1, a bead for 2, a bead for 3, for 4, and so on, monotonously onwards to infinity. But a far more intricate and mysterious structure reveals itself when you thread your beads a different way, with 2 leading the way.

Start with a bead for the number 2 and then follow it with a bead for the number 4, a bead for number 6 and so on. You end up with a necklace made of all the even numbers. Number 3 is not contained in this necklace, so you start a new one with a bead for 3, then a bead for 6, a bead for 9, and so on, until you have all the multiples of 3. The two necklaces will intertwine as they share some beads: 6, 12, 18 and the other multiples of $2 \times 3 = 6$. The next number not contained in our first two necklaces, 5, gives us a third necklace, containing 5 itself and all its multiples. This necklace intertwines with the first one at multiples of $10 = 2 \times 5$, with the second one at multiples of $15 = 3 \times 5$ and with both at multiples of $30 = 2 \times 3 \times 5$.

What emerges is an elegant net of the natural numbers. The beads that appear at the beginning of an individual necklace are the *prime numbers*: they have no factors other than themselves and 1. The number 2 has the honour, not only of being the first prime, but also of being the only even prime. Every other bead is in the interior of the net; it's a product of prime numbers.

Our web represents one of the oldest and most fundamental results in mathematics, appropriately called the *Fundamental Theorem of Arithmetic*: every whole number greater than 1 is either a prime or a product of primes. It was proved around 300 BC by none other than Euclid in his famous book *The Elements*. The theorem also says that there isn't any choice about the primes involved. For example, the only way to get 20 is to multiply two 2s and a 5. Each natural number is built from a unique set of prime numbers.

This turns the primes into the atoms of arithmetic. Just as any molecule is built from a unique combination of the elements described in the periodic table, any natural number is built from a unique combination of prime numbers.

Safety in numbers

Mathematicians have played with, and marvelled at, these atoms of arithmetic for thousands of years, but their power goes far beyond the theoretical. Primes help protect your cash and privacy every time you use the internet.

The secret weapon of the primes is that, while it's easy to multiply them, it's hard to do the reverse. Finding the prime factors that make up a large number takes an enormous amount of computing power. One of the most difficult numbers that has, so far, been factored is called RSA-768, a 232-digit number that is the product of two different 116-digit primes. Factoring RSA-768 required two years of computing time on hundreds of computers, but the task would have taken almost 2,000 years on a single one.

RSA-768 = 1230186684530117755130494958384962720772853569595334792197322452151726400507263657518745202199786469389956474942774063845925192557326303453731548268507917026122142913461670429214311602221240479274737794080665351419597459856902143413

=

33478071698956898786044169848212690817704794983713768568912431388982883793878002287614711652531743087737814467999489

×

36746043666799590428244633799627952632279158164343087642676032283815739666511279233373417143396810270092798736308917.

The difficulty of factoring certain large numbers (particularly those that are the product of two large primes) turns primes into mathematical padlocks. Suppose there is a method for encrypting messages that involves a large number, N, and suppose that decrypting the message requires you to know the prime factors of N. For you to send me an encrypted message, for example your bank details, I first take two large primes and multiply them together. That's relatively easy to do and the result will be the number N, which I now send to you, openly and publicly – N is like an open padlock, anyone can snap it shut but it needs a key to open it. You now use the number N to encrypt your message, snapping the padlock shut. Since I already know the factors of N – they are the key to the padlock – it's easy for me to decrypt the message. Anyone else, however, would have to invest an unfeasible amount of time to factorize N and crack the code. This idea forms the basis for the *RSA public key cryptosystem* used widely to protect both your credit card details and your passwords.

To the limits of complexity

There are two obvious ways of undermining RSA. One is to build faster computers, but (unless you invent a quantum computer) that's easily counteracted by code makers simply using larger primes. The other way could be much more devastating: find an entirely new method for factoring numbers, a method so fast that it puts the world's bank details at your fingertips.

The question of whether such a method exists links into one of the hardest open problems in maths. Factoring numbers belongs to a particular class of mathematical problems called *NP*. Problems in *NP* have answers that are easy to check, that is, a computer can verify that the answer is correct in a reasonable amount of time (and computer scientists have a clear idea of what they mean by 'reasonable'). Factoring a number is in *NP*, as it is easy to check whether your factors, once you have them, multiply together to give the original number. Some of the problems in the *NP* class can also be *solved* in a reasonable amount of time – these form their very own sub-class, called *P*. What we don't know is whether there are quick methods to also solve all the other problems in *NP*, including factorization. We don't know whether the *P* class is in fact equal to the *NP* class. This question, known as the *P versus NP problem*, was posed in 1971 and no one has yet managed to answer it.

The difficulty of *P* vs *NP* was recognized by the Clay Mathematics Institute when it listed it as one of the seven Millennium Prize Problems announced in 2000, offering $US 1,000,000 for a proof one way or the other. Not everyone agrees, but most mathematicians seem to think that *P* is not equal to *NP*; this would mean that *NP* problems really are very difficult, and our mathematical padlocks will be safe as long as we have a steady stream of larger and larger primes with which to build them.

Prime puzzles

Luckily we do have an inexhaustible supply of large primes thanks to another result from Euclid's *Elements*: there are infinitely many of them. His proof of this result (see box) was simple and conclusive, but it did not tell us what all these infinitely many primes are. Euclid's result is an example of something that happens quite a lot in mathematics: you may be able to prove that something exists (infinitely many primes in this case) but your proof doesn't actually describe the objects themselves – it's a *non-constructive* proof.

AN INFINITY OF PRIMES

Again, Euclid's proof that there are infinitely many primes is surprisingly simple and elegant. Imagine there are only a finite number of primes and label them p_1 to p_k. Consider the number $E = (p_1 \times p_2 \times \ldots \times p_k) + 1$. Euclid's *Fundamental Theorem of Arithmetic* says that E is a unique product of primes. However, E is not divisible by any of the primes p_1 to p_k (this is because we have added on 1), so either E is itself prime or there must be some other prime number $p_{(k+1)}$ that divides E, which is not part of our original set of primes. The same argument works for any finite collection of primes – no such collection will ever be enough to build all the natural numbers. It follows that there are infinitely many primes. Et voilà – you're a Greek mathematician!

But that might not seem so surprising – if you wanted to show that there are infinitely many things by explicitly writing them down you'd need an infinite amount of time, right? Not quite. If you ask me to give you all the even numbers, and there are infinitely many of those too, I can simply say: the even numbers are those numbers of the form $2n$, where n is a natural number. This way you can easily identify the 100th even number as $2 \times 100 = 200$.

There doesn't seem to be any similar description of the primes. Looking at the first few primes it's not obvious if there is a rule to describe them all. It quickly becomes apparent that they are not equally spaced:

2, 3, 5, 7, 11, 13, 17, 19, 23, 29, 31, 37, 41, 43, 47, 53, 59, 61, 67, 71, 73, 79, 83, 89, 97.

As you look further along the number line, primes seem to become sparser, but they also seem to bunch up in some stretches more than in others. For example, there are nine primes in the 100 numbers before ten million, but only two primes in the 100 numbers after ten million.

Mathematicians have struggled for centuries to find patterns in the primes, with tantalizing results. For example, every now and again primes seem to come in pairs, only a distance two apart. The first few examples are 3 and 5, 5 and 7, 11 and 13, and 29 and 31. Within the 25 primes below 100 you'll find eight such twin pairs. The largest known twin pair is $3{,}756{,}801{,}695{,}685 \times 2^{666{,}669} - 1$ and $3{,}756{,}801{,}695{,}685 \times 2^{666{,}669} + 1$, both 200,700 digits long, discovered in 2011.

Mathematicians believe that there are in fact infinitely many twin pairs. This so-called *Twin Prime Conjecture* has been around for over 150 years, but so far nobody has been able to prove it. It's an example of a common phenomenon in number theory. A hypothesis might be easy to make and easy to state, but that does not make it easy to prove.

A particularly famous example of this goes back to the 1740s and a correspondence between the German mathematician Christian Goldbach and the legendary Leonhard Euler (we'll meet him again in the next chapter). Goldbach conjectured that *every even natural number greater than 2 can be written as the sum of two primes*. It's easy to see that this is true for the first few even numbers: $4 = 2 + 2$, $6 = 3 + 3$, $8 = 5 + 3$, $10 = 5 + 5 = 7 + 3$. Computers have checked that it's true for numbers up to at least 4×10^{17} but a general proof remains elusive.

Goldbach's Conjecture has become well-known outside of mathematics, starring in movies, novels and TV shows, often used to mark the genius of the character purporting to have solved it.

Prime patterns

The Twin Prime and Goldbach conjectures may give glimpses of the nature of the primes, but they have given us nothing which spells out exactly what all of them are. And that may be because there is no real pattern in the primes; they are essentially sprinkled randomly along the number line.

One way of getting a handle on this is to ask, not *which* numbers are prime, but *how many* of, say, the first N numbers are prime.

A 14-year-old German boy called Carl Friedrich Gauss considered this question at the end of the 18th century with far-reaching consequences. Gauss nonchalantly scribbled into his logarithm table that the number of primes below N was $\frac{N}{\ln(N)}$ (where ln is something called the natural logarithm of N; we'll explain it more in Chapter e). It was an estimate of course. For the first 100 numbers, the estimate gives $\frac{100}{\ln(100)} = 21.7$. That's not a whole number, which the correct answer needs to be, and it's also a little way off the actual 25 primes below 100. Gauss' estimate was later refined and he thought that its percentage error would become smaller as N got larger. In 1896 Charles de la Vallée-Poussin, a Belgian, and Jacques Hadamard, a Frenchman, independently proved him right. Gauss, unsurprisingly given his early contribution, turned out to be one of the greatest mathematical brains of all time.

The result confirmed by Vallée-Poussin and Hadamard is known as the *Prime Number Theorem*. One of its consequences is bad news for a weary traveller wandering along the number line, collecting primes: there are arbitrarily long gaps between them. Give me any number you like and I can guarantee that there are two consecutive primes whose difference is larger than that number.

The true secret of the primes, however, is concealed within one of the hardest maths problems of all time: the 150-year-old Riemann Hypothesis.

Harmonious primes

The year 1859 was a good one for science. It saw the publication not only of Charles Darwin's *The Origin of Species* but also of a paper on number theory by 33-year-old Bernhard Riemann. Gauss' estimate of the number of primes below a given number, N, remained just that, an estimate. The question was how this guess fluctuated around the real count of primes.

Marcus du Sautoy, a mathematician and a musician, gave us our favourite description of Riemann's deep mathematical insight using the language of music. When you play a note on a violin, say an A, what you hear is very different from the same note played on a tuning fork. The reason is that the violin plays not just the pure note but sounds of other frequencies too, called the harmonics. If you plot the graph of the sound wave for the A from the tuning fork what you get is a perfect regular wave. The corresponding graph for the violin, however, is a much more complicated shape. To get from the regular wave to the waveform for the violin you need to add on the waves representing the harmonics. Indeed, the shape created by any type of sound or noise can be constructed from regular waves (see Chapter τ for more).

Now think of the number of primes below a number N, known as $\pi(N)$, as the full sound of the violin. We don't know the value of $\pi(N)$ for all N – that's what we'd like to know – but we can still say something about the shape it creates: the graph we get when we plot $\pi(N)$ against N. Since there can only be a whole number of primes less than a given number – we can't have, say, 34.6 primes – the graph looks like an uneven staircase, remaining constant until the next prime is encountered and then moving up by 1.

The graph of Gauss' estimate of $\pi(N)$, however, is a smooth curve, indicating that the estimate isn't spot-on. It is like the note played on the tuning fork: if you knew the 'harmonics' of the prime number distribution, then you could add them on to the graph of the estimate and, just as in our musical example, build up the true shape of $\pi(N)$.

Riemann's deep insight was that these harmonics were encoded by a mathematical beast that has become known as the *Riemann zeta function*. Unlike Riemann's estimate of $\pi(N)$, the graph of which is a one-dimensional curve, the Riemann zeta function can be used to describe a two-dimensional landscape, a mathematical mountain range. (Technically speaking, the Riemann zeta function takes complex numbers to complex numbers – we will meet complex numbers in Chapter *i*.)

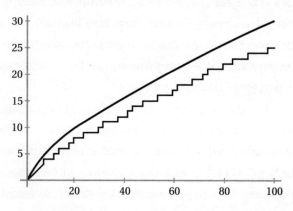

Gauss' estimate compared with the true value of $\pi(N)$

Riemann realized that the points of this mountain range that lie exactly at sea-level held the key to the prime number distribution: they could be used to produce the harmonics needed to build the true function $\pi(N)$ from his estimate. There were infinitely many of these sea-level points, but where were they? In a truly jagged mountain range they could be anywhere, but when plotting some examples Riemann saw an amazing pattern emerge. The sea-level points, at least those that were of interest, all seemed to lie on a straight north–south line, they had exactly the same east–west coordinates. This would mean that the harmonics of the prime distribution are perfectly balanced – none of them dominates the prime number 'sound' – and it would explain why we don't see any dominant patterns in the distribution of the primes.

Riemann couldn't prove that *all* sea-level points, more technically known as *non-trivial zeros*, lie on this magic north–south line, but he conjectured that they did. This conjecture has become known as the Riemann Hypothesis. It has been checked – and found to be true – for ten trillion zeros. But for mathematicians this isn't good enough. Only a complete proof will do.

The Riemann Hypothesis is one of the most important and difficult open problems in mathematics, some would argue it is *the* most important and difficult one. When the British mathematician G.H. Hardy was about to set out on a potentially dangerous sea voyage in the 1920s he sent a note to a friend claiming to have proved it. That way, Hardy thought, God would not allow him to die.

Immortality, if only in name, will certainly be bestowed on whoever manages to finally crack the hypothesis. It featured on a famous list of 23 problems, presented by the mathematician David Hilbert in 1900 as the problems to dominate 20th-century mathematics. A hundred years later, in 2000, the unsolved hypothesis was included in the Clay Mathematics Institute's seven Millennium Problems.

Seven years after making his famous hypothesis, Bernhard Riemann died of tuberculosis at the age of only 39. An over-efficient housekeeper destroyed many of his mathematical notes – whether a proof of his hypothesis was hidden within these we will never know.

Prime quantum chaos

Mathematicians have tried many different strategies in their battle to prove the Riemann Hypothesis since it was first proposed in 1859. One less obvious approach began with the mathematicians David Hilbert (who had provided the list of 23 problems) and George Polya, who independently conjectured that there might be some physical reason why the Riemann Hypothesis might be true. Although this conjecture, now known as the Hilbert–Polya Conjecture, was never published, it is still thought to be the most promising approach to proving the Riemann Hypothesis.

Polya explained, in a letter in 1982 to the mathematician Andrew Odlyzko, how his assertion came about. In 1914 he was spending time in Germany and was working with Edmund Landau, a mathematician who made great contributions to our understanding of primes. Landau asked Polya one day: 'You know some physics. Do you know a physical reason why the Riemann Hypothesis should be true?' Polya's suggestion was essentially that if the Riemann zeta function could be linked to some physical system, say a vibrating system where the zeros of the function were the frequencies of the vibrations, then certain physical constraints would be equivalent to the Riemann Hypothesis being true.

But what could that physical system be? A good place to look is at the very smallest scales of physics, at the *quantum systems* – those involving electrons, photons and the other fundamental particles that make up our Universe. Just like the zeros of the Riemann zeta function, the possible energy levels of these systems can be identified with points on the two-dimensional plane (they are coordinates of points in the *complex plane*, which we'll find out about in Chapter *i*). If mathematicians or physicists could find a quantum system that links the energy levels with the nontrivial zeros of the Riemann Zeta function in the right way, they'd have immediately proved that all the zeros lie on Riemann's magic line.

In the 1980s mathematicians discovered such a connection. The quantum systems involved behave chaotically when looked at from afar using the physics of Newton. Chaotic systems, such as our weather, are very sensitive to small changes, famously described by the butterfly effect we met in Chapter $\sqrt{2}$. The behaviour of objects in a chaotic system, say clouds in a weather system, are incredibly hard to predict, since tiny variations can lead to vastly different outcomes. But sometimes objects will move into a repeated cycle. Mathematicians showed that, when looked at quantum mechanically, this large-scale cyclic behaviour was connected with the quantum energy levels of the system by a formula that was remarkably similar to the formula connecting the zeros of the Riemann zeta function and the primes: the Riemann zeros were like the quantum energy levels and the primes were the periodic cycles.

This connection has had a deep impact in both number theory and physics. Back in 1918 G.H. Hardy (he of the sea voyage) and a colleague, J.E. Littlewood, had attempted to understand how the landscape produced by the Riemann zeta function varied as you moved along the magic line that was conjectured to contain all the non-trivial zeros: how did the *values* of the Riemann zeta function change? They had tried to measure the variability of the Riemann zeta function using a hierarchy of numbers called the *moments* of the function.

They calculated the first moment (which is analogous to the average, or *mean*, of the values of the function) in 1918, and eight years later they managed to calculate the next number in the hierarchy. Progress then stalled until the 1990s when the next two moments were calculated, leading to a sequence of numbers: 1, 2, 42, 24,024. But at this point mathematicians became stuck. Their methods were never going to provide them with any more numbers in this sequence and the movement of the Riemann zeta function along the magic line remained in shadow.

It was the connection to quantum chaos that came to the rescue. By applying the mathematical methods used to describe quantum chaotic systems to the Riemann zeta function physicists derived a formula that would give not just the next moment in the sequence, but every single moment describing the distribution of the values along the critical line. This formula agreed with the moments already discovered and shed new light on the Riemann zeta function, giving mathematicians more ammunition in their battle to prove the Riemann Hypothesis.

What is certain is that to solve the Riemann Hypothesis, as well as any of the other prime number mysteries, we will need to draw on the full power of mathematics, from theoretical physics all the way to algebra and geometry. Calculus will be in there too, of course, and this is embodied by one of the most beautiful numbers of mathematics: the number *e*.

e Naturally!

When Google floated on the stock market in 2004 it aimed to raise $2,718,281,828. That seems like a very odd choice – why that strange amount? Google famously celebrates the maths, science and technology that has enabled it to exist, and this financial target was a chance to recognize a very special number: they were aiming to raise *e* billion dollars, rounded to the nearest dollar.

The number *e* is one of the most important in mathematics, perhaps only trumped by π in its ubiquity. But it is far less famous than some of its mathematical cousins, probably because it is much harder to define. Its tricky definition also explains its ubiquity: *e* often bubbles up when mathematicians push their calculations to the limit. And every time *e* appeared in some complex mathematical machinery, things suddenly became a lot easier.

e is for interest

The first mathematician to catch sight of *e* was the 17th-century Swiss mathematician Jacob Bernoulli (1654–1705), who was part of a famous mathematical dynasty that produced eight outstanding mathematicians in just three generations. Among many other things Bernoulli was interested in infinite sequences of numbers. One example is as relevant today as it was in Bernoulli's time: compound interest.

Suppose you deposit £100,000 with a bank at a yearly interest rate of 4%. How much will you have after two years? It's tempting to guess that the interest after two years would be 8% of £100,000, that's £8,000, so your total amount is £108,000. But of course that's wrong, which is good news for you as a saver (but bad news if you are a borrower!). After a year the interest will be 4% of £100,000, that's £4,000, so the total amount after one year is £104,000. After another year the interest comes to 4% of that amount, which is £4160, so the total amount after two years is £104,000 + £4160 = £108,160.

The way your money accumulates isn't entirely straightforward, however you can describe it with a relatively simple mathematical expression. After one year you have

$$£100,000 + 0.04 \times £100,000 = £100,000 \times (1 + 0.04)$$

in the bank. After two years the interest on the first year's total is

$$(£100,000 \times (1 + 0.04)) \times 0.04.$$

Adding this to the money you already accumulated, which is £100,000 × (1 + 0.04), you get a total of

$$£100,000 \times (1 + 0.04) + (£100,000 \times (1 + 0.04)) \times 0.04.$$

This is equal to

$$£100,000 \times (1 + 0.04) \times (1 + 0.04) = £100,000 \times (1 + 0.04)^2.$$

Repeating a similar calculation will show that after three years you have

$$£100,000 \times (1 + 0.04)^3$$

and after four years you have

$$£100,000 \times (1 + 0.04)^4.$$

This might lead to suspect that after n years you would have

$$£100,000 \times (1 + 0.04)^n$$

and you would be correct. You can be even more general. Write P for the initial amount deposited and r for the interest rate (here we think of r not as a percentage but as a fraction of 100, for example a 2% rate would give us $r = 0.02$). Then after n years you'd have

$$P \times (1 + r)^n.$$

Keeping compounding

So far, so reasonably straightforward, only in reality many banks don't add interest once a year but more frequently, say quarterly or monthly. How does this affect the formula? To find out, let's follow Bernoulli's example and make the numbers a little easier to deal with. Suppose that you borrow £1 at an (admittedly unrealistic) interest rate of 100%, compounded four times a year. Then this means that the bank will add 25% of interest four times a year.

By the same reasoning as above, with an initial deposit of £1, this means that after a year you have

$$£1 \times \left(1 + \frac{1}{4}\right)^4 = £2.44140625$$

With monthly compounding the amount you have at the end of a year is

$$£1 \times \left(1 + \frac{1}{12}\right)^{12} = £2.613035281.$$

Daily compounding gives

$$£1 \times \left(1 + \frac{1}{365}\right)^{365} = £2.714567455$$

and hourly compounding gives

$$£1 \times \left(1 + \frac{1}{365 \times 24}\right)^{(365 \times 24)} = £2.718120712.$$

As you compound more and more frequently the interest you earn in a year increases but it seems that the total amount after a year never gets larger than £2.72. In a paper published in 1690 Bernoulli showed that as you compound interest more and more frequently (as n gets larger and larger) the amount you have after a year, $£\left(1 + \frac{1}{n}\right)^n$, gets arbitrarily close to, but never exceeds a particular number that we know today as e. Although he was only able to approximate this number to lie between 2 and 3, Bernoulli was the first person to recognize e as something special: the *limit* of $\left(1 + \frac{1}{n}\right)^n$ as n tends towards infinity.

Once we know about e, things suddenly get much easier. For any initial amount, £P, and for any interest, r (r is between 0 for 0% and 1 for 100%), as you compound more and more frequently the amount you have after a year gets arbitrarily close to £Pe^r.

We can even say you would have exactly this amount if you compounded the interest *continuously* at every instant in time.

But why would anyone want to do that?

Suppose that after a year of earning interest at 4% compounded monthly on your initial deposit of £100,000 you transfer to another bank with 5% interest compounded quarterly. Then after two years you have

$$£100,000 \times \left(1 + \frac{0.04}{12}\right)^{12} \times \left(1 + \frac{0.05}{4}\right)^{4}.$$

That's not a very nice expression. It will get worse as you juggle your finances transferring investments (or loans) from place to place. For every investment or loan period with different conditions another bracket enters the fray. There is no way of simplifying the expression; when you multiply out the brackets you will get terms such as

$$\left(\frac{0.04}{12}\right)^{12} \times \left(\frac{0.05}{4}\right)^{4}$$

which can't be made any simpler because the rules for dealing with powers or indices say nothing about multiplying a number raised to some power by another number raised to some power. (See the following box for a reminder of the *power rules*.)

POWER RULES

To investigate the arithmetic of numbers raised to a power (called *exponentiation*), let's leave e behind for a moment and look at the slightly less intimidating number 2.

Raising 2 to the power a, where a is a natural number, means multiplying 2 by itself a times. For $a = 3$, for example, we write

$$2^3 = 2 \times 2 \times 2 = 8.$$

This leads to some straightforward rules for exponentiation. For example, 2^a multiplied by 2^b means multiplying 2 by itself $a + b$ times, so

$$2^a \times 2^b = 2^{(a+b)}.$$

Now suppose that a is greater than b and divide 2^a by 2^b. Since a is greater than b you can cancel all the 2s in the denominator (on the bottom) of this expression leaving you with exactly $a - b$ 2s multiplied together:

$$\frac{2^a}{2^b} = 2^{(a-b)}.$$

What if you raise 2^a to the power b? Then you are multiplying 2^a by itself b times:

$$(2^a)^b = 2^{(a \times b)}.$$

But what should you make of raising 2 to some fractional power, such as $\frac{1}{2}$? Following the rules above you should get

$$2^{\frac{1}{2}} \times 2^{\frac{1}{2}} = 2^{\left(\frac{1}{2} + \frac{1}{2}\right)} = 2^1 = 2.$$

Thus, $2^{\frac{1}{2}}$ should be the number that when multiplied by itself gives 2. This is of course the number we met in the last chapter:

$$2^{\frac{1}{2}} = \sqrt{2}.$$

Similarly $2^{\frac{1}{3}}$ is defined to be the number that when multiplied by itself 3 times gives you 2 (it's the cube root of 2), $2^{\frac{1}{4}}$ the number that when multiplied by itself 4 times gives you 2 (it's the fourth root of 2), and, more generally, $2^{\frac{1}{n}}$ is defined to be the number that when multiplied by itself n times gives you 2 (it's the nth root of 2).

To raise 2 to some fractional power that does not have a 1 in the numerator (the top of the fraction), for example $2^{\frac{2}{3}}$, you simply observe that

$$2^{\frac{2}{3}} = 2^{\left(\frac{1}{3} \times 2\right)}$$

which by the rules above should be equal to $\left(2^{\frac{1}{3}}\right)^2$. That's simply the cube root of 2, squared.

More generally for natural numbers a and b we have

$$2^{\frac{a}{b}} = \left(2^{\frac{1}{b}}\right)^a.$$

We can now also make sense of negative powers, such as 2^{-a}. By the rules above

$$2^a \times 2^{-a} = 2^{(a-a)} = 2^0 = 1.$$

So 2^{-a} should be the number that when multiplied by 2^a gives you 1. In other words

$$2^{-a} = \frac{1}{2}^a$$

Simple!!

This doesn't just work for the number 2; you can define negative and fractional powers this way for any number at all, including our new friend, the number e.

But now suppose that both loans compound interest continuously. Then after the first year you have

$$£100{,}000e^{0.04}.$$

Applying the formula again means that after you have been with the new bank for a year, you have

$$£100{,}000e^{0.04}e^{0.05}.$$

Now the rules for exponentiation do apply. They tell us that you can simply add the powers to get

$$£100{,}000e^{(0.04\,+\,0.05)} = £100{,}000e^{0.09}.$$

This is the total amount you'll have after two years. The maths becomes a whole lot simpler using continuous compounding and this is why financial institutions use it for many of their more complex financial instruments, such as options and derivatives.

e is for Euler

Although Bernoulli only approximated the value of *e*, estimating it to lie between 2 and 3, it is the thinking behind this process that is fundamental. The central idea is that an infinite sequence of numbers, say the value of your savings as you compound more and more frequently, can approach a limit. Numbers in the sequence get closer and closer to that limit; in fact you can get arbitrarily close to it by moving far enough along the sequence. Bernoulli's definition of *e* in the late 17th century was the first time that a constant had been defined as the limit of an infinite sequence.

But it was another Swiss mathematician, the prolific Leonhard Euler, whom we already met in Chapter 2, who provided the first accurate estimation of *e*. In his book, *Introductio in analysis infinitorum*, published in 1748, he defined the *exponential function*, which assigns to each number *x* the value of e^x. But he defined this function in terms of a limit: e^x is the limit as *n* goes to infinity of

$$\left(1 + \frac{x}{n}\right)^n.$$

If you plug $x = 1$ into this expression you get the limit as *n* goes to infinity of

$$\left(1 + \frac{1}{n}\right)^n$$

which, as we have seen above, is exactly equal to *e*. You can check that this definition works for any other number *x* using your calculator: for large numbers *n* the value of

$$\left(1 + \frac{x}{n}\right)^n$$

will be very close to that of e^x.

But Euler noticed another intriguing fact, also involving infinity. Consider the sum

$$1 + \frac{1}{1} + \frac{1}{2 \times 1} + \frac{1}{3 \times 2 \times 1} + \frac{1}{4 \times 3 \times 2 \times 1} = 2.70833\ldots$$

It exhibits a pattern which you can easily continue. The next term to add would be

$$\frac{1}{5 \times 4 \times 3 \times 2 \times 1},$$

followed by

$$\frac{1}{6 \times 5 \times 4 \times 3 \times 2 \times 1},$$

and so on. As you add more and more terms in this way, the resulting sums approach a limit and this limit is exactly the number e. Thus we can write

$$e = 1 + \frac{1}{1} + \frac{1}{2 \times 1} + \frac{1}{3 \times 2 \times 1} + \frac{1}{4 \times 3 \times 2 \times 1} + \cdots$$

As before, you can play the same game for any number x. The infinite sum

$$1 + \frac{x}{1} + \frac{x^2}{2 \times 1} + \frac{x^3}{3 \times 2 \times 1} + \frac{x^4}{4 \times 3 \times 2 \times 1} + \cdots$$

is equal to e^x.

It is thought that Euler used this alternative description of e (e^x when $x = 1$) to compute the value of e to 23 decimal places:

2.71828182845904523536028.

Euler also proved that e is irrational as its continued fraction is infinite, like those of $\sqrt{2}$ and ϕ that we met in earlier chapters. And he was the first person to refer to the constant by the letter it has come to be known by: e. Some have suggested the choice of e by Euler was self-referential, but it is far more likely that it stood for 'exponential' or was just the next letter of the alphabet that was not already being used to represent some other value in Euler's work.

e is for Napier

Euler's tour de force about e in his *Introductio* has led many to refer to e as *Euler's constant*. But if e is going to belong to someone, perhaps it should be the English lord and gifted amateur mathematician John Napier (1550–1670), in whose work e made its very first shadowy appearance. Working decades before Bernoulli first caught sight of e, Napier was looking for ways of simplifying calculations involving very large numbers, which arose from the thriving sciences of astronomy and navigation.

Multiplying two large numbers becomes a lot easier if you can use powers. Take, for example, a million and a billion. Once you realize that a million is 10^6 and a billion is 10^9, multiplying them becomes easy as you simply add the powers (using the power rules from the box above):

$$10^6 \times 10^9 = 10^{6+9} = 10^{15}.$$

To exploit the idea to the full you first need to know how to represent one number as a power of another, in this case as a power of 10. This leads to the idea of a *logarithm*. Rather than asking 'what number do I get if I raise 10 to the power of 2',

$$10^2 = ?$$

and getting the answer 100, you ask 'to what power do I need to raise 10 to get 100?'

$$10^? = 100.$$

The answer in this example is 2, called the *logarithm of 100 to base 10* because to get 100 you need to raise 10 to the power of 2. More generally, the logarithm of a number x to *base a* is the number to whose power you must raise a to get x as the result:

$$x = a^{\log x}.$$

In his paper *The description of the wonderful canon of logarithms*, published in 1614, Napier produced what are essentially logarithmic tables relating numbers to their logarithm, but with a twist. A rather obscure geometrical argument had led him to construct his logarithms to the base

$$\left(1 - \frac{1}{10^7}\right)^{10^7}.$$

If you look at this closely you will notice that this expression is very similar to one we already met. It's

$$\left(1 + \tfrac{x}{n}\right)^n,$$

with the x replaced by -1 and the n replaced by 10^7. From Euler's limit definition of e^x we know that this number is very close to e^{-1}. And from the rules of powers we know that e^{-1} is equal to $\tfrac{1}{e}$. Thus, Napier had unwittingly compiled tables giving the logarithm to base $\tfrac{1}{e}$, without ever even having heard of e!

Today we know the logarithm to the base e – the number to whose power you must raise e to get x as the result – as the *natural logarithm* of x, and write it as $\ln(x)$ or $\ln x$:

$$x = e^{\ln x}.$$

e for infinitesimal

Whenever we talk about the number e there is no way of getting around the concept of a limit, but the applications of this concept go much, much further. It's fundamental whenever we want to study how things change.

An example is the speed of a car: speed is distance divided by time. If you have driven 100 miles in two hours then you know that your average speed during your journey was 50 miles per hour. But of course that's only an average; the exact speed with which you were travelling at a given moment in time is likely to be different. It may have been zero if you were stopped at a traffic light, or 100 miles per hour if you were belting down the motorway. Your exact speed at a given time is the *instantaneous rate of change* of distance with respect to time.

To catch such a fleeting quantity you can use a sequence of approximations. Suppose you want to know at what speed you travelled exactly 60 minutes into the journey. A first approximation of this is your average speed throughout the whole journey. You get a better idea by computing the average speed you travelled at in the period between 55 and 65 minutes into the journey. An even better approximation comes from the average speed you travelled at between 56 and 64 minutes. Considering the average speed between 57 and 63 minutes gives you something even more accurate. And so on. Your exact speed at 60 minutes is the limit of average speeds computed for infinitely many time intervals that get shorter and shorter, honing in on 60.

Speed is not the only example of a rate of change you come across in daily life. Acceleration is the rate of change of speed with respect to unit time, growth is the rate of change of size with respect to time, cooling or warming are the rates of change of temperature with respect to time, and so on. The world is full of change, and, as we have just seen, if that change is continuous, then the concept of a limit is essential in getting a mathematical handle on it.

Although the Greeks grasped at that concept of a limit over 2,000 years ago (we'll find out more in Chapter τ), it wasn't until the 17th century that the mathematical handle was fully developed in the shape of what is now known as calculus. We owe it mostly, but by no means exclusively, to two men: Isaac Newton (1642–1727) of Britain and Wilhelm Gottfried von Leibniz (1645–1716) of Germany. Their methods made it possible to calculate reliably with slippery *infinitesimal* quantities: quantities that are so small they can't be measured, but you can still do maths with. Calculus enables us to calculate with rates of change and limiting processes which come up not only in mathematics itself, but also in engineering, biology, astronomy and even statistics. The scope for real-world applications is literally limitless.

What wasn't quite so limitless was the generosity of Newton and Leibniz. Their invention was followed by a bitter dispute between the two men and their followers about who invented the calculus first. Newton claimed to have come up with the central ideas of his theory in 1666, at the tender age of 23, but did not fully publish his work until 1687. Leibniz started working on his version of calculus in 1674, and published his first paper on the topic in 1684.

Nobody doubted that Newton had got there first, but the question was whether Leibniz had developed his theory independently or whether he'd come across manuscripts of Newton's work that provided the vital spark. There is some evidence that he did, and there's also evidence that he retrospectively tinkered with his own manuscripts and falsified dates on them, casting some doubt on his good faith. On the other hand, a report published by the Royal Society in 1713, and finding in Newton's favour, had been orchestrated by none other than Newton himself. Neither of the two men came away squeaky clean. Today the consensus seems to be that they both independently discovered calculus. But in some sense Leibniz won out – the notation he invented to express his calculus was so useable that we still employ it today.

e is for exponential

Calculus also brings us back to *e* and one of its most beautiful features. The function e^x comes with its very own curve: simply mark the values of e^x on a vertical axis and plot them against x measured on a horizontal axis

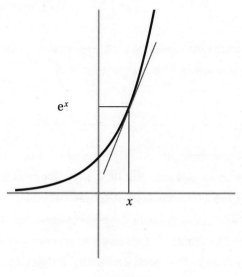

Graph of e^x

The resulting curve becomes steeper as x grows larger, illustrating a phenomenon we often come across in real life: exponential growth. So how fast does e^x grow? How steep is its slope at a given point x? You can measure the slope of a curve by comparing how much it changes in the vertical direction with how much it changes in the horizontal direction: slope is a rate of change and you can work it out using calculus. It turns out that the slope of the curve for e^x at a point x is exactly the value of e^x at that point – the function e^x gives the slope of its own curve!

As you'd expect, this makes finding the slope of any curve involving the exponential function much easier. For example, any variable, y, that changes at a rate proportional to its size (like our example of continuously compounding interest above) can be described by a function of the form

$$y = ce^{kx}$$

for some constant numbers c and k. The slope of this curve (the rate at which y changes with x) is just ky.

e is natural

All these facts are about maths, but it turns out that e has another trick up its sleeve. Nature is full of spirals. There is a spiral in a nautilus shell, many insects trace a spiral when they home in on a light source and distant galaxies have spiral arms. The number e is believed to be intimately related to these natural spirals – and it's also related to the golden spiral we met in Chapter ϕ.

To see why and how, consider two perpendicular axes again, one horizontal and one vertical. Write O for the point where the two axes meet: that's going to be the centre of our spiral. Now imagine putting your thumb down on O, keeping it fixed, and rotating the part of the horizontal axis that lies to the right of O round and round in an anticlockwise direction. Also imagine that there is a pencil attached to the axis, which at the start is at distance 1cm from O. As the axis sweeps round and round, the pencil moves further and further away from O tracing out a spiral. It does so at a particular rate: at any given moment the distance from the pencil to O is exactly equal to e^{α}, where α is the angle you have turned through, measured in radians (you will find out more about this way of measuring angles in Chapter τ). As you turn round and round, through a full turn, two full turns, and so on, the angle α, and hence the distance e^α, gets larger and larger, producing a spiral which grows outwards at a rate that depends on e.

You can also continue the spiral inwards by rotating your line segment in the clockwise direction. Again start with the pencil attached at distance 1cm from O and the line in the horizontal position. As you begin to move it round and round in the clockwise direction, your pencil moves inwards towards O. At any moment its distance from O is $\frac{1}{e^\beta}$, where β is the angle you have turned through. As β gets larger and larger, through a full clockwise turn and beyond, $\frac{1}{e^\beta}$ gets smaller and smaller, so your pencil spirals in on O without ever quite getting there.

This kind of spiral, called logarithmic because of its connection to e, was first discovered by one of our usual suspects, Jacob Bernoulli, and he named it his *spira mirabilis* – the miraculous spiral. It has a number of beautiful features. For example, the distance along the spiral from any point on it to the point O is always finite, even though the spiral winds around O an infinite number of times.

Logarithmic spiral

But the feature that seems to have fascinated Bernoulli most is that the spiral is *self-similar*: you can zoom in or out of the picture above by as much as you like, the spiral you see will look exactly the same as the one you started with, although it may appear to have been rotated.

It turns out that there are many more spirals with this feature, but each is of the same form. Every point on it lies on a line that makes some angle, α, with the horizontal axis and is at distance $ae^{b\alpha}$ from O, where a and b are constants. Our example above has $a = b = 1$. The golden spiral that we met earlier corresponds to $b = 2 \times \frac{\ln(\phi)}{\pi} \approx 0.3$. All of these are logarithmic spirals. And it has been suggested that many of the spirals we see in nature are of this logarithmic form.

Bernoulli was so fascinated by this self-similarity that he planned to have a logarithmic spiral engraved on his tombstone, along with the words (in Latin) 'Although changed, I shall arise the same.' Unfortunately though, the wrong type of spiral, an Archimedean one whose loops are an equal distance apart, ended up adorning his grave. Poor Bernoulli, he must have turned (logarithmically) in his grave. Perhaps he should have gone for a simpler shape, the trusty old triangle.

3 It takes three

One of the neatest rules you learn at school is that the angles of a triangle always add up to 180 degrees. Always. Without exception, whether it is an acute triangle (all the angles are less than 90 degrees) or an obtuse triangle (one of the angles is greater than 90 degrees) triangle, a right-angled triangle, an equilateral triangle (all sides and angles are equal), an isosceles triangle (two sides and two angles are equal) or a scalene triangle (all sides and angles are different).

But – you might want to sit down for this – your maths teacher was lying to you. The angles of a triangle do not always add up to 180 degrees. And the proof of this has been literally under our noses the whole time.

We naturally think in terms of Euclidean geometry, the geometry of the flat plane, which Euclid specified around 300 BC. This makes sense since we draw on flat pieces of paper. But it's in stark contrast to the ground we actually stand on, which is only approximately flat. We live on the roughly spherical surface of the Earth and here conventional Euclidean geometry doesn't work.

The rule that the angles of a triangle add to 180 degrees is a version of Euclid's fifth axiom, which we met in Chapter 1. Its original version was a little more complicated:

If a line segment intersects two straight lines forming two interior angles on the same side that sum to less than two right angles, then the two lines, if extended indefinitely, meet on that side on which the angles sum to less than two right angles.

This is illustrated in the image to the left. The two lines meet a third forming two angles that are less than 90 degrees each, which means, in Euclid's flat geometry, they meet.

If the two lines fail to behave like this, if they don't meet, then they are *parallel*. Euclid's fifth axiom tells us that two lines are parallel if they both meet a third one in a right angle, and this is why it is also called the *parallel postulate*. With a bit of extra work you can show that it is equivalent to saying that the angles in a triangle add up to 180 degrees, at least as long as you are working on the flat plane.

Bulging circles

But now suppose you are standing on the equator between two lines of longitude. The three lines you are looking at, the equator and the two longitudinal lines, are examples of *great circles*: they wrap around the fattest part of the Earth, achieving a maximal diameter – no other circle drawn on the surface of the Earth can be bigger. Great circles also give you paths of shortest distance: the shortest route between any two points on the Earth is along the unique great circle that contains them both. Great circles are to *spherical geometry* what straight lines are to Euclidean geometry, they are paths of shortest distance, and they should therefore obey the same laws.

But here is the surprise: if you squat down and measure the angle each line of longitude makes with the equator, you will find that both angles are right angles. Yet, like all lines of longitude, these intersect at the North and South poles: they are not parallel. What we have here is a surprisingly familiar counterexample to the parallel postulate.

A triangle on the sphere is formed by arcs of great circles. But it bulges outwards and its angles add up to *more* than 180 degrees. How much more depends on the size of the triangle. The sum of the angles in a small triangle will only be a whisker above 180 degrees because for small regions the surface is approximately flat. But as your triangle gets larger and larger, say a triangle connecting London (UK), Münster (Germany) and Perth (Australia), the sum of the angles increases (see image below).

The mismatch between spherical and Euclidean geometry becomes particularly apparent when you look at maps which are, after all, attempts to depict a sphere on a flat piece of paper. Any flat map of the Earth is distorted in some way (as we'll see in Chapter

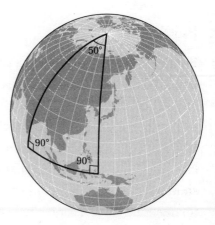

60). The most commonly used, called the *Mercator projection*, is fairly accurate around the equator but grossly exaggerates the distances and sizes nearer the two poles. Greenland seems vastly larger than Africa when in reality it is around 14 times smaller.

Strange new worlds

Once you get used to it, spherical geometry is still fairly natural: you can visualize it, you can draw diagrams of it, and if all else fails you can draw pictures on an orange. There is another type of geometry, however, that is much more exotic. It is called *hyperbolic geometry* and in it angles in a triangle add up to *less* than 180 degrees.

Hyperbolic geometry grew out of mathematicians' desire to eliminate Euclid's fifth axiom altogether. Compared with the other four axioms, which contained simple statements such as 'a straight line can be drawn through any two points', the fifth seemed much too complicated. The idea was to try to deduce it from the others – to show that if the first four axioms hold then the fifth automatically holds as well.

A proof by contradiction, which we met in Chapter $\sqrt{2}$, seems a natural way of doing this. Suppose that on some very strange surface Euclid's fifth axiom doesn't hold (for example, angles in a triangle add up to less than 180 degrees) but that the other four axioms do. Then show that this surface is an absurdity that cannot possibly exist. Trying to prove this was a pastime that seems to have driven some people to distraction. In a letter to his son Janos (written more than 2,000 years after Euclid) the Hungarian maths teacher Farkas Bolyai wrote:

For God's sake, I beseech you, give it up. Fear it no less than sensual passions and because it, too, may take all your time, and deprive you of your health, peace of mind and happiness in life.

But like many others Janos Bolyai (1802–1860) persevered. Try as they might, the strange geometries that mathematicians constructed remained perfectly reasonable. The hyperbolic geometry that eventually emerged is attributed mainly to Bolyai, the legendary Carl Friedrich Gauss (whom we met in Chapter 2) and Nikolai Lobachevsky (1792–1856), who worked on the subject independently.

Exotic vegetables

So what does a surface that exhibits this strange geometry look like? It is hard to visualize, but to get an idea of what it is like think of a kale leaf which gets more and more crinkly towards the edge. To get from a point *A* to a point *B* a little bug living on the leaf has to climb over all those crinkles, preferably along a path of shortest distance. Now imagine flattening the kale leaf, hopefully without squashing the bug, and smoothing it out so that it forms a round disc (you might have to imagine it being made of playdough). The paths of shortest distance on the leaf do not necessarily turn into straight lines on the flattened version of the leaf. Instead, they may turn into different types of curves. This is very similar to the way that some straight lines of the shortest paths around the Earth, great circles, turn into curves on any flat map: for example, if you fly from the UK to Canada, the shortest route will appear as a curved arc on a flat map.

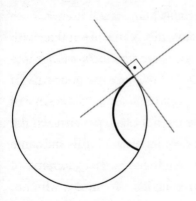

All of the hyperbolic plane fits into the disc bounded by the thin black circle. The role of straight lines in this geometry is played by arcs of circles that meet the boundary in right angles, like the thick black arc shown here.

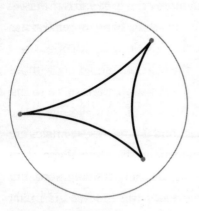

Hyperbolic triangle

The *hyperbolic plane* is infinite in extent, so it's not like a kale leaf, but it is also very crinkly and, by loose analogy, can also be 'flattened' and made to fit into a round disc. A beautiful description of such a flattened map of the hyperbolic plane is named after the French mathematician Henri Poincaré (1854–1912). As in the kale leaf example, paths of shortest distance in this flat Poincaré disc don't run along straight lines. Instead they run along arcs of circles that meet the boundary of the disc at right angles.

Being the paths of shortest distance, these arcs are the 'straight lines' of the Poincaré disc. The sides of hyperbolic triangles are formed by segments of such arcs. This makes the triangles look more pinched at their corners, giving an angle sum of less than 180 degrees.

But how can we say that the hyperbolic plane is infinite when we have just managed to squeeze it into a disc? It turns out that, with this new way of thinking about distances, lengths and areas become more and more stretched as you move towards the boundary of the disc. The distortion becomes so extreme towards the edge of the disc that a hyperbolic being living in it could never reach the boundary – this boundary is infinitely far away. This is the same problem we had with our flat maps of the spherical geometry of the Earth – any flat map of a non-Euclidean geometry, whether hyperbolic or spherical, will always be distorted.

There is a similar model for three-dimensional hyperbolic geometry and even for dimensions beyond that. No wonder that Bolyai exclaimed in a letter to his father, 'I have discovered things so wonderful that I was astounded . . . From out of nothing I have created a strange new world.' Decades later it became clear that hyperbolic geometry is not just a fanciful dream: it is exactly what is needed for the special theory of relativity, developed by Einstein in 1905. We'll find out more about Einstein's theory of relativity in Chapter 4.

Triangular strength

But no matter which geometry you find yourself in, triangles are the first truly interesting two-dimensional shapes: a single point gives you nothing much, two points define a line segment, but three points give you a triangle. One reason triangles are so familiar is that they are all around us. We instinctively use them as they are the simplest shape to build: leaning a stick against a tree creates the beginnings of a simple, triangular shelter.

But triangles have another advantage: they are also the strongest shape. If you join four sticks together at the corners the structure can flex this way and that: no matter the length of your sticks, they can form infinitely many four-sided shapes. These shapes can collapse under the weight of gravity alone, let alone bear any additional weight.

If you connect three sticks at the corners, however, they form a rigid shape; their three lengths define a unique triangle. The rectangle built with four sticks is not rigid, but if you add a diagonal cross-brace your structure is finally solid, thanks to the triangles created.

This simplicity and strength explains why triangular structures, such as the efficient temporary lean-tos thought to have been built by Australian Aborigines for over 20,000 years, were among the earliest built by humans. But triangles are still fundamental in architecture and construction today. Cranes, bridges, pylons and bracing all rely on triangular elements for their strength. And triangles have also become a fundamental part in creating the aesthetics of modern architecture.

Atoms of architecture

The form of 30 St Mary Axe, better known as the Gherkin, is now an established part of the London skyline. Its curved tapered shape stands in sharp contrast to the straight forms that surround it. Its other distinctively curved neighbour is the elegant dome of St Paul's Cathedral built over 300 years earlier. But unlike the curved pieces of masonry of St Paul's almost every piece of glass that makes up the Gherkin's curved form is flat – every one apart from a single curved piece that sits at the very top capping the structure. A similar effect can be seen in the triangulated curves of the new roof of Kings Cross station and the Royal Courtyard of the British Museum. Their three-dimensional forms are described mathematically as computer models, which can be resolved as triangular meshes using mathematical techniques.

The techniques used to triangulate a surface can vary but a simple example is a geodesic dome, built with geodesic lines (the 'straight lines' of a curved surface, such as the great circles on the Earth) that intersect to form triangles. The architect Buckminster Fuller became intrigued by these in the 1960s, entranced by their strength and efficiency. Approximating the shape of a sphere, geodesic domes make efficient use of space, since a sphere encloses the largest volume for a given surface area. And being made of triangles, geodesic domes are also strong.

To approximate a sphere or dome, start with an icosahedron, a three-dimensional shape made up of 20 equilateral triangular faces. It's possible to find an icosahedron which sits exactly inside your sphere, all its corners just touching the sphere's surface. Now no one will be fooled that this is a sphere: the relatively large chunky faces of the icosahedron are like a badly pixelated picture. But we can increase the resolution by using

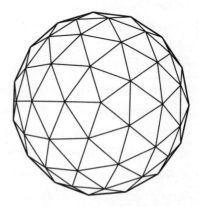

80-faced approximation of a sphere

smaller triangles and many more of them. Divide each of the edges of the icosahedron in half and join these halfway points so each triangular face is now made up of four smaller triangles. You can push these midpoints out until they touch the sphere, giving an object with 80 triangular faces (no longer equilateral triangles) and our flat-faced form is looking distinctly more curvy.

You can carry on this process as long as you like until you achieve your desired level of accuracy for the approximation of the curved surface by your flat triangles.

Triangular monsters

Similar processes can be used to triangulate any surface, approximating its shape with a net of flat triangles, whether it's an architectural form in the real world or the skin of a virtual monster designed to frighten a cinema audience.

Many computer-generated images start off as a triangular mesh of a surface of a three-dimensional object. Just as for architectural models, the mesh must be stored in the computer, often as a list of the positions of the vertices (corners of the triangles) in 3D space and a description of how they combine to make up the triangles. The computer can then shade each of the triangular faces depending on their orientation to the virtual light source. It calculates the orientation of each triangle as a *normal vector*, an arrow pointing perpendicularly out of the flat face, and the angle this normal vector makes with a line from the light source. This angle determines the amount of shading. If the face is flat onto the light source, so the angle the normal vector makes with the line from the light source is close to zero, the triangle is barely shaded at all. Then as the angle increases, and the triangle faces further and further away from the light source, the triangle is shaded darker and darker. These calculations are done using standard techniques from an area of maths called linear algebra. They produce a lifelike play of light and shadow across the surface.

Just as pixels determine the resolution of a flat image, the size and number of triangles in the 3D model determine how realistic it appears. To save computing power you can concentrate the number of triangles (or if appropriate other shapes used in this step) on parts of the model where more detail is needed, such as around the eyes and facial features of a character, and use fewer triangles where less detail is needed, such as on the flat of a flank or back.

The technique can produce stunningly realistic images: you don't even realize you are looking at a shape made out of triangles unless you zoom right in on a little area.

Atoms of geometry

We know that we can triangulate the vast majority of two-dimensional surfaces because of mathematical techniques developed over the last century. Tibor Radó (1895–1965) was a Hungarian soldier who began studying mathematics when he was taken prisoner by the Russians in the First World War. He was taught by fellow prisoner and research mathematician Eduard Helly (1884–1943). After Radó's escape from the Siberian prison camp, helped by the Yupiks (Siberian Eskimos) as he made his way through the Arctic regions of Russia, he became a professional mathematician. In 1925 he showed that it is possible to triangulate all the two-dimensional surfaces you're ever likely to come across.

Such a triangulation tells us quite a bit about the nature of a surface. You might remember from school having to count the corners (or vertices, V), edges (E) and faces (F) of various objects, like cubes, pyramids and prisms, to check that they satisfy the formula

$$V - E + F = 2.$$

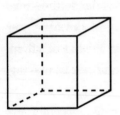 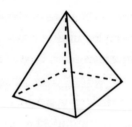 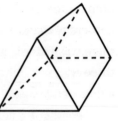

A cube has
8 vertices, 12 edges
and 6 faces, so
$V - E + F = 8 - 12 + 6 = 2$

A pyramid has
5 vertices, 8 edges
and 5 faces, so
$V - E + F = 5 - 8 + 5 = 2$

A prism has
6 vertices, 9 edges
and 5 faces, so
$V - E + F = 6 - 9 + 5 = 2$

This is known as *Euler's polyhedron formula*, after Leonhard Euler whom we met in the last chapter. The result is true of all three-dimensional shapes with flat polygonal faces (and no holes through the shape), whether a pyramid, a cube, a tetrahedron or the geodesic dome that we saw earlier. You can prove this fact by breaking the polygonal faces up into triangles, so the result essentially hinges on a triangulation of your shape.

Thanks to Radó's result we know that any finite surface can be triangulated with a finite number of triangles, giving us a collection of vertices (V), edges (E) and faces (F). The number

$$X = V - E + F$$

is called the *Euler characteristic* of the surface. It's amazing to see that the Euler characteristic doesn't change as you add more triangles to a triangulation. For example, start with the pyramid made up of four triangular faces, also called a *tetrahedron*. It has six edges and four vertices (so $V - E + F = 4 - 6 + 4 = 2$). Now pick the point at the centre of one of the faces and draw lines to the three corners of the face. This gives you three faces in place of the one there was before, giving a total of six faces. You have also gained three edges, making a total of nine edges, and one vertex, making a total of five vertices. The Euler characteristic is

$$X = V - E + F = 5 - 9 + 6 = 2.$$

It has remained unchanged.

Stretch or squeeze, but don't tear

The reason why all these objects – cubes, prisms, pyramids or our deformed pyramid – have the same Euler characteristic is that they are the same in a certain mathematical sense. Two objects are *topologically* equivalent if you can bend, stretch or shrink one into the other without cutting or tearing the object. All the shapes that we have looked at so far are topologically equivalent to the sphere. Simply smooth out their edges and bulge out their faces to make them round. Anything that can be deformed into a sphere in a similar way, including a deflated football or a knobbly potato, has an Euler characteristic of 2.

An example of a shape that has a different Euler characteristic is the doughnut, or *torus* as it is mathematically known. You can make a torus from a flat piece of paper by first joining two opposite edges to make a cylinder, and then curving the cylinder around to join the ends (the other pair of opposite edges in our original piece of paper). So you can triangulate a torus just by triangulating the flat piece of paper, indicating which edges and corners are joined together. The triangulation below has 9 vertices, 27 edges and 18 faces, making the Euler characteristic for the torus

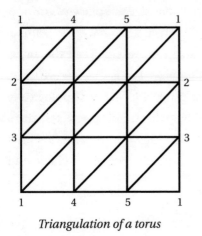

Triangulation of a torus

$$X = 9 - 27 + 18 = 0.$$

And indeed the torus is topologically different from a sphere. To turn a sphere into a doughnut you'd have to make a hole, something not allowed in topology. A coffee cup, however, is topologically the same as the doughnut. Dent one side of your doughnut, making the dent bigger and bigger until it resembles the coffee cup and shrink the rest of the doughnut to become the cup's handle. And as any triangulation of a coffee cup will tell you, its Euler characteristic is also 0.

All this raises an interesting question: is the Euler characteristic enough to classify all surfaces topologically? Not quite. An example is the famous Möbius strip. You can make a Möbius strip by putting a twist in a rectangular strip of paper and joining the ends together. Not only does this join one side of the paper with the other, making the shape one-sided, it also has the effect of joining one edge of the paper with the other, making the edge a single looped circle.

The triangulation to the right, with 5 vertices, 10 edges and 5 faces, gives an Euler characteristic of $\chi = 5 - 10 + 5 = 0$, the same as the torus. So while

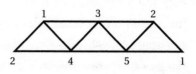

Triangulation of a Möbius strip

two surfaces that are topologically equivalent have the same Euler characteristic, having the same Euler characteristic is not quite enough to guarantee that two shapes are topologically the same. To completely pin down the topology of a surface you need a little more information, for example whether it is one-sided, like the Möbius strip, or whether it has an inside and an outside like the torus.

Topology exemplifies one of the things that mathematicians like to do best: to distil the essence of an object without worrying about its minor details. Another thing they like to do is generalize. When it comes to geometry and topology this means venturing up into higher dimensions. But how is this actually possible? You'll find out more in the next chapter.

4 In another dimension

There is something calming about four. There are four corners to a square, four seasons in a year, four is enough to form two teams and, rather neatly, 4 is both 2 + 2 and 2 × 2. It's a friendly, democratic number.

However, there is one context in which four seems to lead us to a dead end. We live in a three-dimensional world and we cannot imagine one with four. There are three directions to move in: left–right, forward–backward, up–down, and that's it. Any talk of four dimensions smacks of sci-fi.

Think about this again, however, and you realize that all depends on what you mean by dimension. We say that space is three-dimensional because, to locate a point in space, to tell you exactly where I am, I need three bits of information: how far north–south, how far east–west and how far up–down. But sometimes three is not enough: if I want to arrange a meeting with you, I not only need to tell you where, I also need to tell you when. If the number of dimensions is the number of bits of information needed to specify something, then this means that we live in a four-dimensional world, made up of three spatial dimensions and one for time. Moreover, if you are an air traffic controller monitoring, say, ten planes, you are dealing with a 31-dimensional space: three numbers to locate each of the ten planes and one for time. Once you let go of the idea that a dimension is something you should be able to see, any number of dimensions becomes possible.

Mathematicians have known this for a long time and happily talk about 4-, 5- and even n-dimensional spaces where n can be any positive whole number. They have even managed to transfer geometrical concepts into these higher dimensions, despite the fact that we cannot visualize them.

Roundness in the flat

Let's start with a circle drawn on a piece of paper. Trying to describe this circle to someone who can't see is quite hard, until you realize what defines the essence of circleness. A circle is what you get when you take a piece of rope, attach it to a pole in the ground and then walk around that pole keeping the rope taut at all times. As you are walking around the pole you are always at the same distance from it – the length of your rope. Putting it into more mathematical terms, a circle is the collection of points (in a two-dimensional plane) that are all at the same distance from a given point. The given point is the centre of your circle and the distance is its radius.

The plane the circle lives in has two dimensions: we need two bits of information to locate a point within it. Let's take a reference point, called the *origin*, and locate a point in the plane by stipulating how far you have to move in the horizontal direction and how far you have to move in the vertical direction to get to the point. Moving horizontally and to the right is counted using positive numbers and moving to the left is counted using negative numbers; similarly moving vertically up is given by positive numbers and vertically down by negative ones. So the point (12, 4) is the location you get to by moving a distance 12 horizontally to the right and then a distance 4 vertically up. The point (–3, –15) is where you get to when moving a distance of 3 horizontally to the left and then a distance 15 vertically down.

Now think of a circle centred on the origin with a radius of 1. What are the coordinates of each point on that circle? Suppose you draw the straight line from the origin to a point, P, on your circle, then drop down vertically to the horizontal axis and add a horizontal line taking you back to the origin. What you have is a right-angled triangle with the radius of the circle opposite the right angle. The length of the horizontal side of your triangle, call it x, is also the first coordinate of your point, P. The length of the vertical side, y, is its second coordinate (see figure below). From Pythagoras' theorem (see Chapter √2) you know that

$$x^2 + y^2 = 1.$$

This is true for any point P on the circle. And if a point isn't on the circle then this isn't true. In other words, the circle consists of all those points whose coordinates (x, y) satisfy the equation above (we'll find out more about coordinates in Chapter x). By the same reasoning a circle with radius 2 is given by the equation

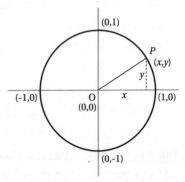

Graph of the circle $x^2 + y^2 = 1$

$$x^2 + y^2 = 4.$$

You can work out for yourself that a circle centred on the point (12, 4) with radius 2 is given by

$$(x - 12)^2 + (y - 4)^2 = 4,$$

and more generally that the equation

$$(x - a)^2 + (y - b)^2 = r^2$$

describes a circle centred on the point (a, b) with radius r. There is a beautiful synergy between geometry and algebra.

Higher-dimensional roundness

With a bit of work you can show that this also works in three dimensions. A sphere is the collection of points that are all at the same distance from a given point. Each point now has three coordinates (x, y, z), as there is extra dimension to move in, and a sphere centred on the origin with radius 1 is given by

$$x^2 + y^2 + z^2 = 1.$$

A sphere centred on the point (a, b, c) with radius r is given by

$$(x - a)^2 + (y - b)^2 + (z - c)^2 = r^2.$$

But why stop here? Just as the two-dimensional plane is defined by pairs of numbers, and three-dimensional space by triples, so four-dimensional space is defined by quadruplets of numbers (x, y, z, w). A sphere in four-dimensional space, centred on the origin and with radius 1 is made up of those points whose coordinates satisfy

$$x^2 + y^2 + z^2 + w^2 = 1.$$

And a sphere in four-dimensional space centred on the point (a, b, c, d) with radius r is given by

$$(x - a)^2 + (y - b)^2 + (z - c)^2 \, (w - d)^2 = r^2.$$

You may not be able to visualize this *hypersphere*, as it is called, because it's hard to imagine all of three-dimensional space curving around a point in the fourth dimension. But it's there, defined using algebra. And of course you can go further, defining spheres (and all sorts of other objects besides) in n-dimensional space for any natural number n.

Getting loopy

This is neat, but practically minded people might interject that it is a little too precise. After all, an orange isn't a perfect sphere, yet it is essentially spherical. The same goes for a deflated football. The surfaces of these objects wouldn't be captured by our equations yet you wouldn't describe them as dramatically un-spherical. Geometry, with its precise measurement of distances and angles, is a little too restrictive here, so we need to revert to something we met in the previous chapter: topology.

We saw that triangulations are a great way of getting to grips with topology but there is another way too: it's to use loops. Suppose you draw a loop through a point on a sphere, a knobbly orange or a deflated football. In each of these examples you can easily imagine shrinking that loop down to a point without cutting it. You just pull it tight. The same isn't true for a doughnut, or for its topological twin, a coffee cup. A loop that winds around its hole can't be shrunk down to a point. For a doughnut there are essentially two types of loop: one type that winds a number of times around the hole and another type that winds a number of times through the hole. Any other type of loop is a combination of those two types, winding round and round this way and that.

Just like the Euler characteristic for the triangulated surfaces from the last chapter, the properties of these loops are unchanged when you bend or stretch the surface (they are *topologically invariant*). For example, imagine you have a compact surface (i.e. it has a triangulation with a finite number of triangles) without a boundary (you never hit an edge as you walk around it) and which has an inside and

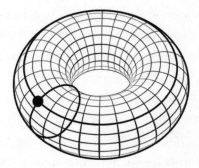

Any loop on the surface of a torus can be described as a combination of loops around and through the hole.

an outside (unlike the Möbius strip we met in Chapter 3). If all loops on this surface can be shrunk to a point, it is topologically equivalent to a sphere.

Moving on up

Now let's do the same in the fourth dimension. In three-dimensional space a surface, such as the surface of a sphere, is a *2-manifold*. Even though it might be curved like the sphere, when you look at it from close up it looks like a plane. The equivalent of a surface in four-dimensional space is called a 3-*manifold*: it's something that when viewed from close-up looks exactly like the ordinary three-dimensional space we are used to. The hypersphere is an example. We can't visualize these 3-manifolds, but by analogy with the two-dimensional situation we are now in the position to formulate a conjecture:

> *Every compact 3-manifold which doesn't have a boundary and in which all loops can be shrunk to a point is topologically equivalent to the hypersphere.*

It was Henri Poincaré who first made this conjecture in 1904 (we first met him in Chapter $\sqrt{2}$). Over the years this conjecture, known as the *Poincaré Conjecture*, became notorious. Many mathematicians tried to prove it; some announced their 'proof' publicly only to find later and to some embarrassment that it contained a fatal flaw. In 2000 the Clay Mathematics Institute turned the conjecture into one of its Millennium Prize Problems (alongside the *P* vs *NP* problem that we met in Chapter 2), offering $1 million to the person who could solve it.

The breakthrough finally came a century after Poincaré first made the conjecture, in a series of papers published online in 2002 and 2003. The author was the mathematician Grigory Perelman and the papers contained a proof, not only of the Poincaré Conjecture, but of the more general *geometrization conjecture* that had been formulated in 1982 by the mathematician William Thurston. After intense scrutiny Perelman's approach was found to be correct. In 2006 Perelman was awarded the Fields medal, one of the highest honours of mathematics, and in 2010 he was awarded the Millennium Prize.

But Perelman's reaction stunned the world. He refused both honour and money, resigned from his post at the Steklov Institute in Moscow and reportedly moved in with his mother. In refusing the Fields medal he was reported as saying: 'If the proof is correct then no other recognition is needed.' Unsurprisingly the mainstream media pounced on the chance to peddle the age-old cliché of the eccentric and reclusive mathematician. On the upside, this is probably the only way a sophisticated mathematical result will ever get coverage in the tabloid newspapers. Whether or not Perelman has resumed his mathematical work is unclear.

What about dimensions greater than 4? If 4 was hard to nail, then surely things can only get worse as we move up the dimensions? Surprisingly that's not the case. The generalized version of the conjecture, regarding manifolds that live in a space of dimension greater than 4, was proved in 1961 (for spheres living in dimensions 6 and greater) and in 1982 (for dimension 5). The fourth dimension really is strange.

Stretching time

All this is abstract maths, but does the fourth dimension really matter in the real world? In 1905 Einstein showed that it does, at least if you are a physicist trying to understand the workings of the Universe. In his special theory of relativity, published in that year, Einstein postulated that the speed with which light travels through empty space (the vacuum) is constant: it appears the same, no matter how fast the source of the light is moving compared to you. This may seem counterintuitive at first. You may have experienced the fleeting union with people travelling on another train as your train moves alongside theirs travelling in the same direction at the same speed, creating an apparently stationary world. In ordinary life, your perception of speed does depend on an object's motion relative to your own.

But Einstein had a reason for making this curious postulate. In 1873 James Clerk Maxwell published his theory of electromagnetism and suggested, correctly, that light itself was electromagnetic radiation. His equations predicted that all electromagnetic radiation, including light, should travel at a universal speed. Maxwell's theory was so impressively accurate that Einstein decided to take its prediction as read, elevating it to a fundamental tenet of his theory.

This had a strange consequence on time. Suppose you and a friend have agreed to measure the passage of time from two different viewpoints. Your friend is travelling on a train and you are standing by the side of the tracks. Time can be measured in all sorts of ways, but you and your friend have decided to use a torch that emits a pulse of light. Your friend fires the pulse vertically upwards the moment the train passes you. The time taken for the pulse to travel to the ceiling of the train is the distance to the ceiling divided by the speed of light (speed = distance / time can be rearranged as time = distance / speed).

The trouble is, however, that the distance from the torch to the ceiling appears longer to you, standing by the side of the track, than it does to your friend on the train because you also observe the horizontal forward motion of the train. To you the light appears to be moving along a diagonal line and that diagonal line is longer than the vertical one (see the figure opposite).

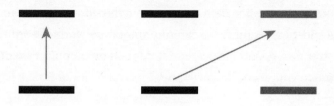

On the left is the distance the light travels as it appears to your friend on the train. On the right is that distance as it appears to you (the train moving from the black position to the grey position between the time when the light pulse was emitted and when it hit the ceiling).

Making the same calculation using the new, longer distance, with the same speed of light gives you a larger value for the time the pulse took to travel to the ceiling. Time as measured by your friend's clock on the train is slower than time as measured by you. The fact is that your different states of motion put you and your friend into different *inertial frames of reference*, as physicists call them. The values you measure in an experiment like this one depend on the inertial frame you are in: in other words, time is relative.

Special relativity showed that time is not absolute: it cannot be separated from motion through space. Einstein's second big idea, general relativity, went further. It showed that motion isn't the only thing that can distort time; gravity can do the same. A clock on the ground ticks slower than a clock on the tenth floor of an office block, because near the surface of the Earth its gravitational pull is stronger. It's only a tiny effect, but it's measurable. This is reflected in the mathematical formalism that describes relativity theory: its space and time coordinates are intertwined. To do physics you must consider a four-dimensional spacetime.

We don't notice this *time dilation* in everyday life because the speeds we are moving at are slow compared with the speed of light, which is a staggering 299,792,458 metres per second. Our smart phones and satnavs, however, must take account of it. The global positioning system that so many of us have come to rely on uses satellites that are moving at great speeds, around 14,000 km/h relative to our position on Earth. Special relativity says that as they are travelling at such speeds the clocks on the GPS satellite would tick more slowly than those on Earth. But of course general relativity says that time is also affected by gravity, with clocks ticking faster the further they are from a massive object. Taking these two effects into consideration the clocks on a GPS satellite would gain 38 microseconds a day compared with clocks on the ground. If the satellites' clocks were not adjusted to take these relativistic effects into account they would cause the GPS calculations to be out by more than 10 kilometres in one day!

10, 11 and counting

Einstein's relativity theory does a great job of describing the world at the scales of planets, stars and galaxies, taking account of the force that governs how these massive objects interact: gravity. But when it comes to very small things, atoms and subatomic particles, we need to turn to the other great success story of the 20th century, quantum mechanics. Like relativity it was developed early on in the century. By the 1950s, and not without considerable mathematical difficulties, physicists had developed a quantum mechanical description of the force of electromagnetism, describing the interaction of light and matter. Their efforts eventually culminated in what is known as the *standard model of particle physics*: a theoretical framework describing the fundamental particles and the forces through which they interact, based on the insights of quantum mechanics. Although a work in progress, the standard model does phenomenally well when tested against reality in experiments.

But despite these successes physicists are not happy. The standard model does not account for gravity and its description of the other forces is fundamentally different from Einstein's description of gravity, both conceptually and mathematically. The thing to do, then, is to bring gravity into the fold, giving it the same mathematical and conceptual treatment as the other forces – to *quantize* gravity. The trouble is that gravity refuses to comply. Any attempt to quantize it gives nonsensical answers, which, if taken at face value, would imply that spacetime should tear itself apart. Finding a unified theory of *quantum gravity* remains the biggest challenge for physics in the 21st century.

There is some hope, however. In the 1980s physicists realized that a bold assumption would get them around the mathematical and conceptual problems that arise from trying to quantize gravity. Rather than thinking of fundamental particles, such as electrons, as tiny little points, we should think of them as tiny little strings. The idea is that all the physics we observe, the fundamental particles and their interactions, result from the vibrations of these little strings, just as the rich experience of a violin concerto results from vibrating violin strings. This so-called *string theory* is described in a mathematical language that is seductively elegant and free from contradictions. Most importantly it incorporates a description of all the fundamental forces, including gravity.

As you might have guessed already, it turns out that in string theory a four-dimensional spacetime is not enough. In order to create all the physics we observe, the little strings need more directions to wriggle in than three spatial dimensions would allow. String theory requires ten dimensions, nine spatial ones and one for time, in order to work. And there is a generalization of string theory, called *M-theory*, which requires eleven.

As we have seen, it's no problem to deal with these extra dimensions mathematically, but we still need to square them with real life. Why don't we see them? One idea is that we are stuck in a four-dimensional subworld of a higher-dimensional universe. Imagine a colony of tiny flat bugs trapped forever on the surface of a petri dish. There is a whole third dimension out there, an up-direction, which the bugs will never experience. As far as they are concerned they live in the two-dimensions of the dish and they will never get out. Another idea is that the extra dimensions are rolled up so tightly, they are so small, that our coarse-grained perception just doesn't pick them up. Going back to the bugs, imagine that at every point on the dish there is an extra direction to move in, a third dimension, but that this direction loops back on itself. It is a tiny little circle and anything that ventures along it immediately arrives back where it started. If this circle is many, many times smaller than the bugs, so small they could not possibly see it, they will never be aware of it. In both of these examples the bugs have no need to evolve an ability to perceive the third dimension. Knowing how to move around in two serves them just fine. By analogy, we may not have evolved an awareness of higher dimensions simply because we don't need it.

So when will we know if all this is really true? Not any time soon. The fundamental strings of string theory are so small, predicted to be just 10^{-34} metres long, that current methods are unable to detect them and will remain so for the foreseeable future. Neither will there be any other ways of testing string theory in the lab any time soon. This is why critics regard it as nothing more than a fanciful mathematical dream. There is a little bit of hope for extra dimensions though. Theories that predict them also predict certain kinds of exotic particles. These are similar to known ones, but their masses are much higher. In theory at least, such particles could be detected in particle accelerators like the Large Hadron Collider at CERN. If they are, then this would count as evidence that extra dimensions really exist. It would show that four is really just a gateway to a world much weirder than we could ever have imagined. If this makes your head spin, then you might want to return to something a lot more down to Earth: interior decoration.

5 Trouble with tiles

I f you're thinking of tiling your bathroom there's quite a lot of choice in terms of colours, but not that much when it comes to the shape of your tiles. Your local DIY store will definitely have square and rectangular tiles, perhaps hexagonal ones and if you are very lucky you may even find triangular ones. But if your favourite number is five, you're stumped – you won't find a pentagonal tile.

Five won't fit

The reason is simple. Regular pentagons just don't fit together to tile a bathroom wall, or any other sort of flat surface. That's because the angles in their

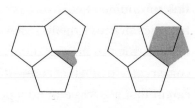

Pentagon tiles

corners are 108 degrees. In a regular tiling several tiles fit around a shared corner point, so their corner angles must add up to a full turn, or 360 degrees. If you fit three pentagons around a point you only get 3 × 108 = 324 degrees, so there is a gap. If you try to fit four of them you get 4 × 108 = 432 degrees, which is more than 360, so they overlap. Even if you offset the pentagons against each other, so the corner of one meets another mid-side, counting angles shows very quickly that things won't fit.

This, incidentally, is also the reason why you won't find *regular* tiles – with sides all the same length and the same angles between them – that have more than six sides. If a regular *polygon* (shape of tile) is suitable to tile the plane (fit together to cover a flat surface with no spaces), then, as we have seen, its angle must divide 360. Since we always have at least three tiles meeting at a corner point, the angle can be no bigger than $\frac{360}{3} = 120$. That's exactly the angle of a regular hexagon, which gives the familiar honeycomb pattern. But, as you can easily see when you draw a few regular polygons, the angle between sides gets larger the more sides there are, so anything that has more than six sides will have an angle greater than 120 degrees. And that's too large.

Just to double check, you *can* make a tiling with equilateral triangles because their corner angle is 60 degrees. Since 360 = 6 × 60, you can fit six of them around a corner. The angle in the corners of a square is of course 90 degrees, and as 4 × 90 = 360 you can fit four squares around a point. And since squares are a lot easier to make than triangles, the vast majority of bathroom tiles are square.

Pentagrams and monsters

But is all hope really lost for fans of five? The pentagon is just one of many shapes that have *5-fold symmetry*. Another one is the five-pointed star, or *pentagram*, and there is also the ten-sided regular *decagon*. Each of these can be rotated around its centre point by a fifth of a turn (72 degrees) and it will look the same after the rotation as it did before. Perhaps you can tile the plane with a combination of tiles that all have 5-fold symmetry?

You can try and you would be in good company. Many great mathematical minds, including the 17th-century genius Johannes Kepler, more famous for his laws of planetary motion, have played around with the *5-fold tiling problem*. But all of them got stuck eventually. In his 1619 book *Harmonices mundi* ('The harmony of the world') Kepler produced a famous patch of tiling using the pentagon, pentagram and the decagon, but he was forced to admit what he called 'monsters': shapes made by gluing two decagons at a side, which break the 5-fold symmetry. Nobody has as yet been able to come up with a tiling consisting of tiles that have 5-fold symmetry that can be continued indefinitely. Neither has anyone been able to prove that such a tiling doesn't exist. So our naive considerations of bathroom decoration have led us straight to an unsolved problem in mathematics.

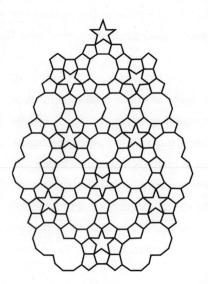

Kepler's tiling

Sweet symmetry

One way of avoiding the 5-fold tiling problem is to tile other types of surfaces. On a sphere twelve regular pentagons fit together neatly, and on the hyperbolic plane, which we met in Chapter 3, you can make a tiling with four regular pentagons fitting together at their corners. Both of these surfaces are curved, so it seems to be flatness that fiveness doesn't get on with.

Another way is to simply let go of our preference of five, or the idea that we should use only one type of tile. The stunning mosaics we see in Islamic art, for example in the Alhambra in Granada or the Topkapı Palace in Istanbul, are made up of all sorts of differently shaped pieces, but they still strike us as highly symmetrical. Even a young child would spot that symmetry immediately, as they would the symmetry of a butterfly or a snowflake magnified under a microscope.

But ask someone, even a grown-up, what symmetry actually is, and the chances are they will hesitate before replying. It takes some thought to come up with the answer: symmetry is immunity to change. The pentagon has 5-fold *rotational symmetry* because when you rotate it by 72 degrees it looks the same as before. A butterfly has *mirror* (or *reflective*) *symmetry* because when you reflect it in the line down its centre it looks the same. And a line of identical houses in a residential street has *translational symmetry*: a giant could shift them all along by the distance of one or more houses, and the street would still look the same.

This makes the circle the ultimate symmetrical object. You can rotate it around its centre through *any* degree and it will look the same. And you can reflect it in *any* line that passes through its centre without changing it. It's a curious fact that the circle's perfect symmetry goes largely unnoticed – if we are asked to name the first symmetrical object that springs to mind we're far more likely to pick a square or a butterfly. Perhaps it is the isolated nature of their symmetries that's striking.

When it comes to people and animals it is often broken symmetry we notice most: a lopsided smile, the nose slightly bent to one side, one eye slightly higher than the other. Some have suggested that symmetry is a prerequisite for beauty. An asymmetric body or face may betray some health or genetic defect that we instinctively flinch away from since it's only the fittest we want to breed with. But on the other hand a lopsided smile can be very sexy, making its owner stand out among bland and uninteresting symmetrical faces. And an asymmetry that's due to a past injury may be attractive to those who prefer their partners battle worn. As far as symmetry and human beauty are concerned, it's probably fair to say that the jury is still out.

Staring at walls

Armed with a proper understanding of symmetry – as immunity to change – we can return to interior decoration. A regular bathroom tiling is an example of what mathematicians call a *wallpaper pattern*. This means exactly the same as it does in ordinary life: it's a pattern that repeats in two directions, based on symmetry. The only difference is that mathematicians don't care whether it is actually made of paper or ceramic tiles. Their approach to understanding such a pattern is to write down all of its symmetries. Forget the roses, teddy bears or other intricate designs that make it up, and concentrate on the transformations that leave it unchanged. These include the familiar reflections, translations and rotations, but also so-called *glide reflections*, which consist of a reflection followed by a translation parallel to the reflection axis. An example of a pattern that is symmetric under a glide reflection is a line of footprints in the sand.

Could a wallpaper pattern have 5-fold rotational symmetry? Given the awkwardness of 5 when it came to individual tiles you might guess that the answer is no and you'd be right. Perhaps surprisingly, given the near infinite wealth of different wallpaper designs, there is a strict limit to the configurations of symmetries you can get: there are only seventeen so-called *wallpaper groups* and none of them involve the number 5.

People seem to have been aware of all seventeen wallpaper patterns for a long time. Famously almost all of the patterns (fourteen at the last count as judged by mathematicians at the International Congress of Mathematicians in Madrid, Spain, in 2006) have been found in the centuries-old decorations of the Alhambra. But it wasn't until 1891 that it was proved that these seventeen patterns are all there are.

Not just a pretty face

The first to prove this result was the Russian mathematician Evgraf Fedorov, in his work *The Symmetry of Regular Systems of Figures*. As well as a mathematician, Fedorov was also a chemist and this was his motivation for the proof. Chemists use the symmetry of individual molecules or groups of molecules in crystals to identify materials and to understand their behaviour.

For example, the V-shaped water molecule remains indistinguishable under four symmetry operations, this time in three-dimensional space, rather than just the two-dimensional plane. There is a 2-fold axis of rotation running down vertically through the oxygen molecule and between the two hydrogen atoms. There are also two

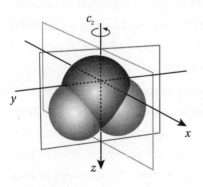

The V-shaped water molecule, H_2O

planes of reflection: one in the plane containing the centres of all three atoms of the molecule, and one passing through the oxygen atom and between the two hydrogen atoms. Together with the identity symmetry operation (doing nothing), these four operations are the only symmetries of a water molecule.

Fedorov wasn't interested in decorations when he exhaustively identified all the wallpaper patterns; he was trying to stipulate all the symmetry groups that were possible in two-dimensional crystals. And just one year later Fedorov, together with the German mathematician Artur Schönflies, listed almost all the 230 symmetry groups possible in three-dimensional crystals.

Scattering conventional wisdom

One way to determine the symmetries of crystals is to use crystallography, in which an X-ray (or more recently a beam of electrons or neutrons) is fired at the crystal. The atoms of the crystal scatter the beam to produce a diffraction pattern. The beam splits to emerge from different gaps between the atoms, the wave fronts either interfering constructively (enhancing each other, creating a bright spot) or destructively (cancelling each other out, creating a dark spot) when they hit the detector. The ordered structure of the atoms in the crystal means that very sharp diffraction patterns are produced, and the symmetries in these patterns stem from the symmetries of the crystal structure.

Crystals are made up of repeating patterns of atoms called *unit cells*. A two-dimensional crystal is a plane tiled by identical two-dimensional unit cells, which is unchanged by translating the plane in two different directions. And we know from above that only 2-fold (rectangular), 3-fold (triangular), 4-fold (square) and 6-fold (hexagonal) rotational symmetries are allowed for such unit cells. A three-dimensional crystal is made up of identical three-dimensional unit cells that have translational symmetry in three different directions. And again, only 2-, 3-, 4- and 6-fold rotational symmetry is possible for these three-dimensional unit cells to perfectly pack together. These symmetries of the unit cell can be apparent to the naked eye, for example in the cubic structure of rock salt or the hexagonal structure of water-ice in snowflakes.

So imagine the surprise of the Israeli chemist Dan Shechtman on the morning of 8 April 1982, when he looked down his electron microscope. He saw a beautifully clear diffraction pattern, indicative of a highly ordered crystal structure. But his diffraction patterns indicated that his crystal had the forbidden 5-fold symmetry.

If this was true, then the crystal could not be made up of a repeating pattern of unit cells with translational symmetry. But what other highly ordered structure was possible? The explanation to Shechtman's mysterious crystal lay in mathematicians' games.

Penrose tiles and the forbidden 5

Mathematicians are quite like naughty children; they are always pushing their boundaries. Unless you put your foot down with a rigorous mathematical proof to show that something is impossible, they won't be stopped from leaning as far as possible over the edge of a known mathematical cliff. In the 1960s mathematicians understood very well how to tile the plane in a pattern that repeated *periodically* (at regular intervals), producing one of the wallpaper groups we've investigated above. Of course this prompted some contrary types to ask if you could tile a plane without translational symmetry, with a pattern that doesn't regularly repeat . . . the hunt for a *non-periodic* tiling was on.

The rules were that you needed a limited set of tiles that you could repeatedly use for your non-repeating pattern. The American mathematician Robert Berger came up with the first winning set of *aperiodic* tiles in 1966, but it consisted of 20,426 differently shaped tiles. Smaller and smaller sets were found over the years until the British mathematician, Roger Penrose, finally came up with the best-known example in the 1970s, a set of just two aperiodic tiles. The *Penrose tiles* consist of one fat and one thin rhombus (a diamond shape), which, when placed according to certain rules, produce an infinitely non-periodic pattern.

But a non-repeating pattern does not necessarily mean a disordered pattern, quite the opposite. In every one of the infinitely many possible tilings with Penrose tiles, there will be finite local areas with 5-fold rotational symmetry. And some of the tilings will have 5-fold symmetry globally as well. There is a three-dimensional version of Penrose's tiles, consisting of one

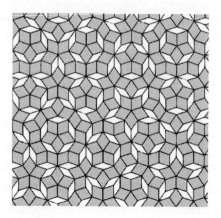

A non-periodic tiling using Penrose tiles

fat and one thin *rhombohedra* (squashed cubes) that also have five-fold symmetry, either locally or globally. And it was such an ordered, non-periodic, three-dimensional structure that caused Shechtman's unusual diffraction patterns – he wasn't looking at a crystal, but at the first identified *quasicrystal*.

Unlike Penrose's discovery, which was immediately celebrated by the mathematical community, Shechtman was initially ridiculed for his results. He was even asked to leave his research group as they felt he was an embarrassment. But as some chemists began to realize that they themselves had seen similar forbidden symmetries in diffraction patterns, which they had written off as mistakes, Shechtman's work was slowly accepted and more and more quasicrystals were discovered. Finally recognizing that Shechtman had discovered a new type of crystal structure, not only did the scientific community rewrite the definition of a crystal, they also awarded Shechtman with the Nobel Prize for Chemistry in 2011. (Perhaps mathematicians' quickness to party is because they know they're unlikely ever to win a Nobel Prize – scandalously there is no Nobel Prize for mathematics. Boo hiss!)

Symmetry making

Symmetry also lies at the heart of physics. As the Physics Nobel laureate P.W. Anderson said: 'It is only slightly overstating the case that physics is the study of symmetry.' One of the greatest realizations in physics, exemplified by Isaac Newton's insights into universal gravitation, was that the laws of physics should be the same everywhere. Whether you conduct your experiments here on Earth, on the dark side of the Moon or, indeed, on the other side of the galaxy, the results can be predicted by the same physical laws. The laws of physics are symmetric under translation.

Such inherent symmetries in the Universe have profound consequences, as the German mathematician Emmy Noether showed in a result published in 1918. Noether was one of the few female mathematicians at the turn of the century. Originally trained as a school teacher, in 1900 she made the unexpected decision to study mathematics at university when she was eighteen years old. This was at a time when the university system was not particularly welcoming to women. They were allowed to sit in on lectures if they had permission from the lecturer (fortunately for Noether her father was a mathematics professor and many lecturers were his friends) but they were not awarded degrees and were not allowed to hold academic positions. Thankfully Noether's passion for mathematics overcame all these obstacles and she made significant contributions to the development of *ring theory*, an area of abstract algebra.

In 1915 Noether was invited to the University of Göttingen by Felix Klein and David Hilbert, although it took the university four years (after Klein and Hilbert waged a campaign on her behalf) to finally deign to pay her a meagre salary. In the meantime, Einstein visited the university in 1915 to present his soon to be complete theory of general relativity. One of the remaining problems in the theory was that energy did not seem to be conserved. (Conservation of energy is one of several well-established *conservation laws*: in an isolated system, energy can be neither created nor destroyed, which explains why any plans for a perpetual motion machine are doomed to failure.) Noether set to work on the problem and, in solving it, proved a far-reaching result: any conserved quantity – energy, momentum (the mass of an object multiplied by its velocity) or the many others discovered since – is associated with an underlying symmetry in the physics.

Noether's theorem revealed the deep connection between the symmetry of physical systems and the laws of conservation. Energy is conserved because the laws of physics are symmetric with a shift in time: an apple falls from a tree in exactly the same way as it did when it (allegedly) landed on Newton's head and inspired his theory of universal gravitation. Similarly, conservation of momentum is a consequence of the symmetry under translation. And conservation of *angular momentum* (the momentum of a spinning object, rather than one moving in a straight line) stems from physics that is not changed by how a system is oriented in space: you could rewrite global coordinates so that direct north was pointing towards our office in Cambridge, and the laws of physics would operate in exactly the same way as they did before.

Noether's theorem has become part of the toolkit that theoretical physicists use every day. But it is also vital to discovering new physics, revealing new conserved quantities, such as a property of particles called *colour charge*, and underpinning the theory of *supersymmetry*, which is currently being tested at the Large Hadron Collider in Switzerland. Noether's result, connecting conservation laws with the symmetries of physics, is central to modern physics.

Symmetry breaking

Also central to modern physics is the absence of symmetry, as symmetry breaking is intimately involved in the emergence of order. A glass of liquid water is highly symmetric. The molecules in the water are oriented every which way, making it impossible to tell the difference if you turn the glass by any degree – and, as we saw above, such immunity to change is exactly what symmetry is. But rotate a block of ice and it is a very different story. In most commonly found ice the molecules are arranged in a hexagonal lattice and if you rotate the crystal by anything other than a multiple of 60 degrees you'll easily spot the change in the ordered arrangement of molecules. Therefore, the ice is less symmetric than the water.

The emergence of order in a physical system is essentially a point where symmetry is broken, whether it is in freezing water, magnetizing metals (in which electrons align in the same direction) or a material losing all electrical resistance, becoming *superconducting*.

One of the most exciting recent events in physics – the discovery of telltale signs of the Higgs boson in the Large Hadron Collider in 2012 – is also down to symmetry breaking. In the 1960s theoretical physicists began to suspect that two of the fundamental forces – the electromagnetic force that governs the interactions of electrons and photons, and the weak nuclear force that is responsible for radioactive decay – were in fact different aspects of a unified electroweak force. The electromagnetic force is mediated by charge-carrying particles called *photons* – when two particles interact it is by exchanging photons. The symmetry of the mathematical equations indicated that there must be force-carrying particles for the weak force that were playing the same role as the photons for the electromagnetic force. This symmetry predicted the existence of particles called W and Z *bosons*, which were experimentally observed in the 1980s when experimental physics finally caught up with the theory.

There was a problem, however: the symmetry decreed that these particles should have no mass, just like their electromagnetic equivalent, the photon. But in order to explain the relative weakness of the weak force these bosons had to have a mass. This broken symmetry was explained by the *Higgs mechanism*, developed in the 1960s and named after one of its inventors, Peter Higgs. A by-product of this mechanism was the prediction of another type of particle, called the *Higgs boson*, which was finally glimpsed at CERN, the European Organization for Nuclear Research, nearly forty years after it had been predicted.

Considerations to do with symmetry have taken us quite a long way – from tiling our bathrooms to the emergence of order in the Universe. There is one part of life, however, we are only now beginning to understand: our very own social systems. And it turns out that the number 6 has a lot to teach us here.

6 Bees do it, we do it . . .

Unlike pentagons, the six-sided hexagon has no trouble tiling the plane. Hexagonal tiles are available from your nearest hardware store, and even nature loves to use them. Bees build a honeycomb of hexagonal wax cells to store pollen and honey and to house the bee larvae.

Honeycombs and the strange nature of mathematical proof

As well as being delicious, honeycomb illustrates a fascinating mathematical idea: a honeycomb is the most efficient way to divide a plane into equal-sized areas, in that it uses the least amount of wax. We have known about this since at least AD 300 when the Greek mathematician Pappus of Alexandria introduced the fifth book in his series *The Mathematical Collection* with a charming diversion called *On the Sagacity of Bees*. People have believed for millennia that this *Honeycomb Conjecture* is true, but it wasn't until 1999 that mathematician Thomas Hales confirmed it with a mathematical proof.

Hales' proof of the Honeycomb Conjecture came shortly after his groundbreaking proof in 1998 of the 400-year-old *Kepler Conjecture*, which essentially said that the most efficient way to stack oranges, in the sense that it leaves the smallest gaps possible, was in the familiar pyramid you see at the greengrocers. The question had been posed in the 16th century by Sir Walter Raleigh (who also brought

Europe the potato) to his assistant Thomas Harriot, although Raleigh was thinking of canon balls rather than oranges. Harriot eventually forwarded the question to Johannes Kepler, whom we have already met in the context of tilings in Chapter 5, and it now bears his name.

Hales' proof of Kepler's Conjecture was groundbreaking because a significant proportion of the proof was done using a computer. Although this was still unusual in 1998 it wasn't entirely new: ever since the computer-assisted proof in 1976 of the *four-colour theorem* (which states that you need no more than four colours to colour any flat map so that no two adjacent countries or regions are shaded in the same colour), the mathematical community has had to come to terms with this drastic change in the nature of mathematical proof. Mathematicians had always prided themselves in needing nothing but a pen, paper and their brain to prove a theorem, but now a machine was doing some of the work for them. How could they be sure there wasn't a bug in the code? At the time of Hales' result, the *Annals of Mathematics*, one of the most prestigious journals in mathematics, had been handling computer-assisted proofs for nearly a decade. But Hales' proof was another ballgame altogether: it consisted of 250 pages of traditional mathematical proof and over 3 gigabytes of computer code and data. The *Annals* assembled a team of twelve experts to verify the proof and after four years the best they could say was that they were 99% certain the proof was correct – everywhere they looked it was correct but they had to admit they would never be able to check the whole thing.

The *Annals* and other mathematical journals have had to introduce specific policies to deal with computer-assisted proofs: the human part of the proofs is refereed to the same standard as always but the computer parts are treated more like an experiment – the methods are verified and other mathematicians are encouraged to replicate the results, preferably using different methods.

In 1999, just one year after Hales' mammoth proof of Kepler's Conjecture, Hales turned his attention to the problem of honeycombs. Hales said that after his experience with Kepler's Conjecture he had 'come to expect every theorem to be a monumental effort'. To his surprise, his proof of the Honeycomb Conjecture took only six months to complete (compared with 'years of forced labour' for the proof of Kepler's), was 'just' twenty pages long and required no significant use of computers: 'I felt as if I'd won the lottery.'

As well as being a win for mathematicians, the Honeycomb Conjecture does have an important consequence for bees. The conjecture proves they made a wise decision using the honeycomb structure for their hives: it uses the least amount of wax. And making wax is very expensive for bees: they collectively need to consume about 6 pounds of honey to make one pound of wax, and they'd need to fly the equivalent of 9 times around the world to collect enough pollen to produce that much honey. However, it's unlikely that the reason why bees use a honeycomb structure is because they beat Hales to a proof by millions of years. It's thought they actually build roughly circular cells, which when pressed against each other deform into hexagons, the way soap bubbles form flat sides when they join together.

Six degrees and small worlds

Bees, like us, are highly social animals, but people are not quite as close to each other as bees are. We are separated by geography, by social structures, by our walks of life and our likes and dislikes. But are we really? The number 6 tells a different story: it seems that any two of us are linked by only six degrees of separation in terms of acquaintance. That includes you and the Queen, you and the actor Kevin Bacon and you and the maverick mathematician Paul Erdős.

The idea originated in an experiment conducted in the 1960s by the sociologist Stanley Milgram, who is perhaps better known for another, more sinister, run of experiments designed to understand the horrific actions committed by thousands of 'ordinary' German citizens during the holocaust. Milgram found that a shocking number of volunteers were prepared to administer what they thought were deadly electric shocks to a fake test subject, purely on the grounds that an authority figure had asked them to.

Milgram's small-world experiment was of a far less depressing nature. He wanted to investigate something many of us have experienced: you meet someone far from home and to your surprise it turns out you share a mutual friend or acquaintance. This common experience inspired the *small-world problem*: can you link any two people by a short chain of mutual acquaintances? And how long is such a chain? To investigate this, Milgram randomly selected people from the state of Nebraska and asked them to pass on a letter to a target person in Boston, Massachusetts, a suitable distance away (Milgram was at Harvard, just up the road) via a chain of friends and acquaintances. If they didn't know the target person personally, they were asked to send the letter to someone they knew on a first name basis who they thought would be closer to the target. Although only a small proportion of the letters made it all the way, the average number of links in these chains was 6: and the idea of six degrees of separation was born. The idea was so surprising that it inspired a broadway play (which coined the phrase 'six degrees of separation'), a movie, a TV show and even a charitable social network.

Today we are more aware than ever that our lives are played on and through networks: as well as social networks there are infrastructure networks such as the power grid, water and transportation networks, the physical network of computers and the virtual network of web pages that makes up the internet, even biological networks of neurons in the brain and metabolic processes within our cells. All of these networks are a collection of *nodes* – people, power stations, computers or neurons – connected by *links* – friendships, power lines, wifi or internet cables, and neural connections. And all of these networks appear to exhibit a similar structure. The average distance between two nodes, measured as the number of hops it takes (using links) to get from one node to another, tends to be small. These networks also all tend to have lots of local *clusters*: if two nodes are connected to each other, the other nodes they are connected to tend to be connected too. These two features define what mathematicians call a *small-world network*.

Random reconnections

Why are these small-world networks so ubiquitous? Within social networks you would expect lots of local clusters – friends of friends tend to make friends. At the same time, acquaintances can form in airports or in flings on holiday, loosely linking distant groups. This idea is reflected in a model developed in the late 1990s by the mathematicians Steven Strogatz and Duncan J. Watts. Watts was inspired by his own experience, having moved to Cornell

University in the US from his home in Australia to undertake a PhD with Strogatz. His decision to move across the world suddenly linked together two otherwise distant groups of people, his friends in Australia and his friends in Cornell.

To try to model small-world networks Watts and Strogatz did what mathematicians often do. They started with the simplest setup they could find: They considered a ring of friendships in which each person is connected to a specified small number of people to their left and right along the ring. Such a network has many little friendship clusters – if two people are friends then some of their friends know each other too – but if you choose two people at random, the friendship chain that links one to the other tends to be long.

Watt and Strogatz then started rewiring their neighbourly network, randomly reconnecting some of the links, keeping one end of the link fixed and connecting the other end to a randomly chosen node. Because most of the network isn't in the near neighbourhood of any one node, it's very likely that the rewiring will connect two otherwise disconnected clusters, drastically shortening the path between them. It turned out that only a little bit of random reconnecting sufficed to make the average path between nodes shorter without destroying the local clusters: a few 'shortcuts' between previously distant nodes were enough to make their networks small-world.

The number of hops it takes, on average, to get from one node to another in such a model for a small-world network depends on the total size of the network and the number of nodes that a given node is connected to. In a completely random network, in which each node is connected at random to k others, the average

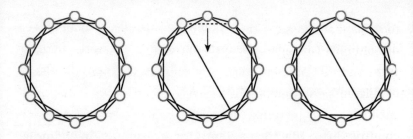

Watts and Strogatz rewired a very ordered network, where nodes were only linked with their nearest neighbours. A link was rewired to a random node in the graph and only a few such rewirings were necessary to create a small world.

distance between two nodes is proportional to $\frac{\ln(N)}{\ln(k)}$, where N is the total number of nodes. Such a network is what you'd get if you took the random rewiring to the extreme. But what Strogatz and Watts found is that you don't have to rewire too many nodes in the initial 'large-world' network before the average path length drops quite rapidly towards this limiting value.

To put this into context, the current world population is around 7 billion. If you discount 15% for people whose social patterns are atypical, because they are babies, too old or unusual in other ways, you are left with 5,950,000,000 people. Assuming that each person has 35 acquaintances on average, we can estimate the average distance between two people in acquaintance terms as

$$\frac{\ln(5{,}950{,}000{,}000)}{\ln(35)} \approx \frac{22.5}{3.55} \approx 6.34.$$

Several years after the discovery of this surprisingly simple mathematical explanation for the small-world phenomenon Watts took his mathematical know-how to the sociology department of Columbia University, where he became a professor. In 2001 Watts and his Columbia colleagues conducted a modern version of Milgram's experiment, this time using the internet rather than snail mail. They recruited over 60,000 people from 166 different countries to reach 18 different target recipients including a professor in the US, a tech consultant in India, and a policeman in Australia. As before they were asked to reach their target by passing on an email (in place of the paper letter) to a friend or acquaintance who they thought was closer to the target. And accounting for the rate of dropouts, this experiment again found that the average number of steps in the email chain was around 6. Studies of online social networks have come up with a slightly different number, presumably due to the ease with which people connect here. Results gave 3.43 for Twitter and 4.75 for Facebook – it seems the internet really does make the world smaller.

Tipping the scales

The average path length and amount of clustering are not the only way you can mathematically describe how a network is connected. The number of links from a node is called the *degree* of the node. Watts and Strogatz's model starts off with everyone having the same number of friends, that is, each node has the same degree. And the degree distribution for the nodes stays the same even with the rewiring. However, when the Hungarian physicists Albert-László Barabási and Réka Albert, working at the university of Notre Dame in the US, started to examine many real-world networks they found a surprising result. While many nodes had similar, relatively small degrees, you could always find nodes with degree at almost any scale, and in particular, there were always a few nodes with a huge number of links – the degree distribution in the networks was *scale-free*.

Barabási and Albert were able to characterize this phenomenon mathematically by plotting the number of nodes of each degree, 1, 2, 3, . . . What they spotted was a *power law*: the number of nodes with k links, $N(k)$, is ak^{-b}, where a and b are some positive constants. Power laws appear repeatedly in mathematics and physics.

Albert and Barabási explained the predominance of scale-free networks by an overarching rule that governs their growth – the rich get richer. When a new node joins a network, it's more likely to make links with nodes that already have a lot of links: you often meet people through friends so you're more likely to befriend someone with a lot of friends already. Over time these already popular nodes gather more and more links until they have a huge degree. Any activity on the network will almost certainly go through these hubs – they govern the behaviour of the network.

It's easy to understand how this happens with the networks of air travel, where the hubs – such as London Heathrow, Beijing Capital and Atlanta – are integral to the network. Any journey that isn't a direct flight will pass through a hub of some degree, and very often through one of the largest. This is such a defining feature of the air travel network that the Federal Aviation Authority (FAA) categorizes all airports in terms of where they fit into the scale of hubs.

And the air travel network also makes it easy to visualize the strengths and weaknesses inherent in a scale-free network. If a random airport shuts down due to bad weather or emergency it will only affect the flights that pass through that airport. It's very likely that such a random failure will affect relatively few flights as most of the airports in the world have very few links ('non-hubs' in the FAA terminology). But if one of the main hubs were to fail the consequences ripple far out across the network, such as when heavy snow essentially closed down Heathrow the weekend before Christmas in 2010, one of the busiest travel periods of the year. Not only did it affect the thousands of passengers travelling through Heathrow itself (including one of us!), but the stranded planes that couldn't travel to Heathrow blocked up other airports, drastically affecting those too. Scale-free networks are very resilient to random failures but extremely vulnerable to the failure of main hubs as these hold the network together.

Why swine flu

We tend to think of social networks as friendly things but when it comes to infectious diseases they can also be deadly – or at least lead to an uncomfortable period bunged up in bed with a runny nose and aching joints.

One of the simplest ways of describing the spread of diseases was developed in the late 1920s and early 1930s by two scientists working in Edinburgh, William Kermack and Anderson McKendrick. Given the 1918 Spanish flu pandemic barely a decade earlier, understanding disease dynamics must have seemed particularly pressing: that particular strand of the H1N1 virus killed between 50 and 100 million people worldwide, that's 3–5% of the entire world population. It's been termed one of the world's deadliest natural disasters and it was certainly the worst flu pandemic the world had ever seen.

Kermack and McKendrick's idea was simple. They divided the population into three classes: people who are infected (called class I), people who are susceptible (class S: they're not infected but can be) and people who have recovered (class R). They then came up with a set of relatively simple mathematical equations that describes the flow of people from one class to another over time. This was called the SIR model. The equations involve two numbers, the transmission rate and the recovery rate of the disease. If you can estimate these from your observations of how the disease affects people then the model will predict if and when the disease will peak or whether it will just dwindle away and die out.

One nice and easy prediction of this basic model is that the course of a disease hinges on one number, the ratio between the transmission rate and the recovery rate. That *basic reproduction number* measures the average number of people a sick person goes on to infect. If it is bigger than 1 at the start of the disease (that is, a sick person infects more than one person on average) then the disease will grow; otherwise it will die out.

The basic reproduction number gives a great way of seeing at a glance how dangerous a disease is. For the 2009 swine flu pandemic it was estimated at 1.3, and as we know that outbreak did turn into a pandemic. For measles it's believed to be a whopping 12. If we assume that people's contact patterns are fairly uniform throughout the population, then this also gives us a good idea of how to plan a vaccination strategy (if you are lucky enough to have a vaccine). Some straightforward maths shows that if the basic reproduction number is R we should vaccinate a proportion $1 - \frac{1}{R}$ of the population; so if $R = 1.3$ then that means a proportion of

$1 - \frac{1}{1.3} = 0.23$, corresponding to 23% of the population. That will drive the basic reproduction number down to an *effective* reproduction number that is less than 1, so you have a good chance of stopping your disease from turning into a pandemic.

But the problem is that people's contact patterns are *not* the same throughout a population and they don't remain constant over time either, so we can't treat the basic reproduction number as a constant quantity. Everyone with children knows this: during term time the little darlings turn into germ-laden disease vehicles. The pattern of the 2009 swine flu pandemic reflected this very clearly: the number of infected people in the UK peaked twice, once in mid July and once in mid-September when children had come back to school after their long summer holiday. The standard SIR model, which assumes that the transmission rate is the same for everyone and stays fixed over time, simply can't predict such a pattern.

This is why epidemiologists are extremely keen to get their hands on information about people's social networks. In some cases this can be tricky. When it comes to sexually transmitted diseases, such as HIV, asking people about their contact patterns is quite intrusive – in fact, in the 1980s Margaret Thatcher's government blocked funding for a survey into people's sexual behaviour for just that reason.

But in other cases it's easier. In the middle of the 2009 swine flu outbreak, scientists from the London School of Hygiene and Tropical Medicine in London launched an online flu survey which invited everyone to regularly report on their health, whether they had been sick or not, and also to provide some basic information about their lifestyle. Over 5,000 people signed up during 2009. The resulting data gave researchers a handle on the type of variability we have already spotted. Scientists Ellen Brooks-Pollock and Ken Eames used the information to create an SIR model made up of two components, one for adults and one for children, that also took into account the different contact patterns in and out of term time. The pattern predicted by their model was impressively close to the real one, showing just how important a little social insight can be.

Ripples of death

Another factor that obviously affects the way a disease spreads, which has changed drastically since Kermack and McKendrick's time, is travel. Data from the Black Death pandemic that hit Europe in the 14th century shows that the disease spread just as we would expect: like ripples in a pond, moving roughly as fast as someone could walk or ride a horse in a day, emanating in concentric circles from places of outbreak. Today our complicated travel networks mean that an infected person hopping on a plane from London to New York is enough to destroy this predictable pattern.

But again there are clever ways to deal with that. In a recent study the physicist Dirk Brockmann simply decided to redefine distance. Rather than using kilometres, or miles, he used the proportion of people travelling between two places to define distance. Despite being geographically distant, according to this new metric, London and New York would be closer than, say, London and the German village of Abtsgmünd, because a higher proportion of travellers leaving London travel to New York than to Abtsgmünd.

Redrawing the network of our towns and cities with this new notion of distance, Brockmann recovered exactly the same ripple pattern for the spread of modern diseases, such as an *E. Coli.* outbreak in Germany in 2011. This result isn't just neat, but also useful: it allows us, for example, to easily pinpoint the place a disease first started, one of the major problems facing epidemiologists.

Brockmann's idea has turned a mess of a pattern into neat round circles. Another example of such a shift in viewpoint has become so iconic most of us aren't even aware of its ingenuity. It's the London Underground map. In terms of geography the tube map is woefully inadequate: something you will soon find out if you try to use it to make your way across London on foot. The distances between stations are grotesquely distorted and in reality the tube lines don't run along as straight as the map would have you believe. For tube travel, though, the map is brilliant, clearly demonstrating which lines are connected to which and where.

The distorted layout was the brainchild of Harry Beck, who worked for the London Underground in 1933 and was familiar with the neat layouts of electronic circuits. Geographically accurate tube maps had stations crowding together in an unintelligible mess in central London to accommodate all those far away stations on the edge. People were scared of taking a journey into the centre and by the 1920s the London Underground was worried about its income. Beck's masterpiece changed not only the fortune of the Underground, but also the attitude of Londoners to their city. It also helps the millions of tourists who visit London each year by making travelling through London easy as pie. Talking of which . . .

ͳ Easy as π?

In 2011 Shigeru Kondo, a Japanese systems engineer, and Alexander Yee, a 23-year-old graduate student, established a new world record: they calculated over ten trillion digits of the number π. With numerous hardware failures and a narrow escape from a major earthquake it took Kondo's modified home computer nearly a year to accomplish the task. At times the poor computer was working so furiously it heated the room to almost 40 degrees centigrade. 'We could dry the laundry immediately, but we had to pay 30,000 yen (about £190) a month for electricity,' Kondo's wife Yukkio told *The Japan Times*.

Why bother trying to do this? Because π is one of the most fundamental numbers in mathematics. It is the ratio between the circumference of a circle and its diameter. It doesn't matter how large or how small the circle is, whether you're on the Earth or on the Moon, its circumference is always equal to π times its diameter. And its area is equal to π times its radius squared. Even better, π is an irrational number. Like e, ϕ and $\sqrt{2}$ it can't be written as a fraction; its decimal expansion is infinite and non-repeating. It's a perfect playground for computer geeks – with such infinite variety the computational challenge never ends.

Kondo and Yee's result is a culmination of over two millennia of mathematical history. The Babylonians, the Egyptians, the Greeks, the Chinese, the Indians, the Arabs and even the Old Testament of the Bible – all had stabs at pinning down π (1 Kings 7:23 if you want to look up the Bible reference). The main problem facing early geometers was that it's hard to measure a curvy shape. The first person to come up with a systematic approach to the challenge was Archimedes (famous for shouting 'Eureka!' in the bath) in the third century BC. He sandwiched a circle between two regular polygons with 96 sides each and thereby arrived at the estimate

$$\frac{223}{71} < \pi < \frac{22}{7}.$$

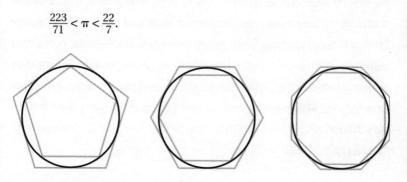

*Estimating the value of π by sandwiching a circle
between two regular polygons*

If you convert these fractions into decimals (and round a little) this gives

$$3.141 < \pi < 3.143,$$

so it only pins down the first three decimal digits of π.

It's clear that you can improve this estimate, making it as accurate as you like, by increasing the number of sides of the polygons involved. Archimedes was actually playing with the concept of a *limit*: the idea that a sequence of numbers (the perimeters of the polygons inscribed in the circle) can grow arbitrarily close to a bound (the circumference of the circle). The Greeks anticipated the invention of calculus by over two thousand years (see Chapter *e*).

Pretty as π

In turn, the age of calculus, with its penchant for infinite processes, gave π its next major boost. During the 17th and 18th centuries mathematicians came up with ways of calculating π that on the face of it have nothing to do with geometry. Instead, they involve infinite sums or products. A beautiful example comes from starting with 1, subtracting $\frac{1}{3}$, then adding $\frac{1}{5}$, subtracting $\frac{1}{7}$, adding $\frac{1}{9}$, and so on. As you keep going, the result of the sum approaches $\frac{\pi}{4}$:

$$1 - \frac{1}{3} + \frac{1}{5} - \frac{1}{7} + \frac{1}{9} - \ldots = \frac{\pi}{4}.$$

This is a truly remarkable result. On the left of the equation you have nothing but arithmetic – the odd numbers, subtraction and addition – and on the right you have π with its deep roots in geometry. The sum carries the name of Wilhelm Leibniz, one of the inventors of calculus (we met him in Chapter *e*), and the Scottish mathematician James Gregory, although it also seems to have been discovered three centuries earlier by the Indian mathematician Madhava of Sangamagrama. Indian mathematicians made quite a few contributions to the type of geometry that involves π which Western histories sometimes forget to mention.

There are other, similarly surprising, expressions for π that are all infinitely long (see the box). These are pretty, but most of them are still quite useless when it comes to calculating the digits of π, especially when you haven't got a calculator. If you add/subtract the first 100 terms in the Madhava–Gregory–Leibniz sum and multiply by 4, the number you get is accurate to only two decimal places, which is disappointing considering the effort involved. By around the middle of the 20th century, only just over 600 digits of π were known, and known correctly, but all this changed with the advent of the computer from the 1950s onwards. During the 1990s new records for calculating π to a huge number of decimal places were established by two Russian brothers, David and Gregory Chudnowski, who had come up with an expression for π that converges quite quickly. It is their formula that underlies the latest record. Retrieving the first 100 digits of π is now just a matter of a Google search:

π = 3.14159 26535 89793 23846 26433 83279 50288 41971 69399 37510 58209 74944 59230 78164 06286 20899 86280 34825. . .

And thanks to Kondo and Yee we also know that its ten trillionth digit is 5.

π SERIES

$$\frac{\pi}{2} = \frac{(2 \times 2 \times 4 \times 4 \times 6 \times 6 \times \ldots)}{(1 \times 3 \times 3 \times 5 \times 5 \times 7 \times \ldots)}$$ (John Wallis 1655)

$$\frac{\pi^2}{6} = 1 + \frac{1}{2^2} + \frac{1}{3^2} + \frac{1}{4^2} + \ldots$$ (Leonhard Euler 1748)

$$\frac{\pi^3}{32} = 1 - \frac{1}{3^3} + \frac{1}{5^3} - \frac{1}{7^3} + \ldots$$

$$\frac{\pi^4}{90} = 1 + \frac{1}{2^4} + \frac{1}{3^4} + \frac{1}{4^4} + \ldots$$

For practical purposes we don't need to know many digits of π. Knowing it to 39 decimal places is enough to calculate the circumference of a circle the size of the known Universe with an error that's smaller than the size of a hydrogen atom. Engineers make do with far less accurate approximations of π. They have to, since no computer or measuring device can cope with its infinitely many digits. π appears in equations involving round objects, as you would expect. But it also appears in many others, particularly those describing waves: radio waves, sound waves, micro waves, water waves, light waves and all sorts of other oscillations that occur in nature and human-made technology. It's an extraordinary stroke of mathematical luck that all wave forms, no matter how complicated, can be represented by a formalism that comes from walking round and round a circle.

The ultimate wave

To see how, imagine the circle drawn on a flat map, so that its centre sits on the equator. To make things easy, take the radius of the circle to be 1. Now start at the circle's eastern-most point and walk around it in an anticlockwise direction. The distance you have covered after walking around half of the circle is

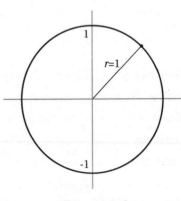

The unit circle

π – that's because the circle has radius 1, so its diameter is 2, making its total circumference equal to 2π. What happened to your north–south coordinate as you walked around the top half of the circle? It started at 0 (since you started at the equator), then increased steadily until it reached a maximum of 1 when you were on top of the circle, and then dropped down to 0 again in a symmetrical fashion as you walked to the western-most point of the circle.

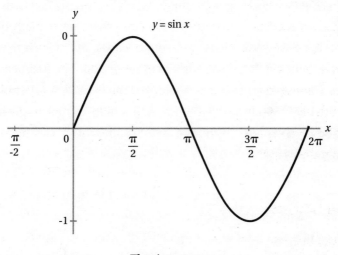

The sine wave

Now as you walk around the bottom half of the circle, the distance walked increases from π to 2π, the north–south coordinate drops down to a minimum and comes back up again to 0. Everything is perfectly symmetrical – points of the circle that lie on the same vertical line have the same distance to the equator. You can make a plot of how the north–south coordinate varies by marking the distance travelled (from 0 to 2π) on a horizontal axis and the corresponding north–south coordinate on a vertical one. What you get is a regular wave shape, starting at 0 going up to the maximum, down to 0 again, then down to the minimum and up again to 0.

If you traverse your circle a second time, the distance walked increases from 2π to 4π; the pattern repeats. The same happens if the distance increases from 4π to 6π then from 6π to 8π, and so on. You end up with an infinitely long, perfectly regular (periodic) wave. The east–west coordinates give a similarly perfect wave, offset against the first one by a distance of $\frac{\pi}{2}$ along the horizontal axis.

The two waves are, respectively, a *sine wave* and a *cosine wave*. If you have heard the terms *sine* and *cosine* before it was probably in your trigonometry lessons, which were all about the relationship between the angles and the sides of triangles. But an angle is nothing but a turn through a fraction of a circle: a 180 degree angle corresponds to half of a circle, a right angle to a quarter of a circle, and so on. Elementary textbook definitions of sine and cosine only work for triangles that have one right angle, so their other two angles can add to no more than 90 degrees – otherwise the total angle sum would exceed 180, which is not allowed for triangles drawn on a plane. The waves we have found above are an extension of this traditional definition to include angles that are greater than 90 degrees.

Decoding messages

At first sight these waves appear too rigid to tell you much about the oscillations we encounter in real life. If you have a sound recorder that displays sound graphically and record yourself singing a single note, what you get will be a lot more jagged. That's because you're not just singing a pure tone, you are also singing lots of harmonics. Similarly, the radio waves sent out by GPS satellites to help you work out where you are don't arrive as clean waves either, they are likely to be corrupted by random interference from other waves.

But what makes sine and cosine waves so powerful is that *any* wave-like signal can be broken up into pure sine and cosine components. We know this because of a result published in 1822 by the French mathematician Jean Baptiste Fourier. Fourier is a great counterexample to the idea that people produce their best mathematics before they are 30. He started out training as a priest, then joined the French revolution, was incarcerated during the Terror but managed to avoid the guillotine, and later travelled to Egypt as scientific advisor to Napoleon's invasion. It was at the age of 54, when he was back in Paris at the Académie des Sciences, that he produced his seminal book *Théorie analytique de la chaleur* (*The Analytic Theory of Heat*).

As the name suggests Fourier came across his result as he was trying to describe how heat flows through a metal plate and realized that it's easier to assume that the heat source behaves in a periodic fashion, just like a wave. Loosely speaking, his important result says that almost any periodic curve can be represented as a (possibly infinite) sum of sine and cosine waves: objects like

$$y = \cos(x) + 4\sin(0.5x) + 0.5\cos(10x),$$

with y plotted on a vertical axis against x on the horizontal one. If your computer has a graphing tool, you can get it to plot this sum to see that you get something a lot more complicated and jagged than a single sine or cosine wave (though still periodic). The Scottish physicist Lord Kelvin hailed Fourier's book a 'great mathematical

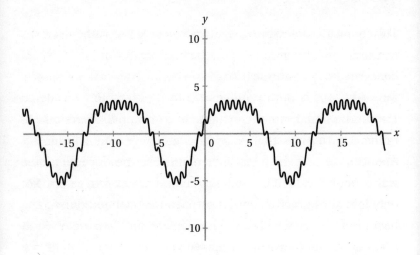

Graph of $y = \cos(x) + 4\sin(0.5x) + 0.5\cos(10x)$

poem' and the analogy fits this result – like composing a poem from words, you can compose complicated shapes from pure sine and cosine waves.

Fourier's result isn't much good to us if we don't know what the constituent components of a complicated wave form are, but remarkably mathematicians have developed a technique, called Fourier analysis, which helps them figure out just that. All types of signals can be decomposed in this way so Fourier analysis has a huge range of applications. It is used to analyse the radio signals from GPS satellites that tell you where you are. It can be used to understand the sound of a real instrument by synthesizing it on a computer, or to clean up a digital sound recording. In medical science it can be used to reconstruct the insides of a person from a CAT scan and in image processing to compress digital images and eliminate defects. The list of applications is almost endless.

Tau – the new π

This connection between π and waves has led some mathematicians to attack several centuries of mathematical convention. They are concerned by the dominance of π = 3.14159. . . over its double, 2π = 6.28318. . . Both numbers could rightfully be considered *the* fundamental circle constant. π is the area of a circle of radius 1 and 2π is the circumference of a circle of radius 1. And 6.28318. . . is the period of repetition of the sine and cosine waves. Yet it is 3.14159. . . that has achieved universal status. Not only is it strewn across countless pages of textbooks stretching back over centuries, it has also appeared on *The Simpsons* and had a feature-length movie made in its honour.

6.28318. . . enjoys no such fame, and some rebels argue that this is not fair. It's the circumference of the circle that is relevant to angle measurement. A quarter of a circle describes a 90 degree angle, a half a circle describes an angle of 180 degrees, and so on. We are more used to dividing the circle into 360 degrees. This is probably a remnant from the Babylonians and it's not clear why they chose 360 over any other number. It may be because the Sun moves across the sky by roughly $\frac{1}{360}$ th of a circle during a day or because their number system had 60 as its base and $6 \times 60 = 360$. Whatever the reason, the subdivision is really rather arbitrary. It makes more sense to define an angle in terms of the distance you need to walk around a circle and that's what mathematicians, physicists and engineers do.

Since angles and waves are used everywhere in maths, and in its applications in science, technology and engineering, rebel mathematicians think that 6.28318. . . should be given its very own

Greek letter. They have chosen τ (tau, pronounced 'taw') and argue that this is what textbooks should use from now on. One supporter, Michael Hartl, has set up a τ manifesto, and the issue even made national news on τ day 2011 – that's 28 June, written 6.28 in the American way, which gives the first three digits of τ.

Whether the campaigners will be able to squash centuries of tradition remains to be seen. After all, when it comes down to it, it really doesn't matter which symbol you use, τ, 2π or a smiley face – it's the concept of the constant that counts. Most non-mathematicians probably could not care less about the π versus τ debate. When you ask them to name a favourite number, they'll probably come up with something far less exotic.

7 Are you feeling lucky?

What is your lucky number? If you said 7 then you are not alone. The number 7 often wins polls of favourite or lucky numbers. Although there are many reasons, rational and irrational, for our love or dread of particular numbers, there is quite a sound basis for the popularity of 7.

Over the centuries we have been playing games of chance – flipping coins and rolling dice – which have honed our intuitive feel for probabilities and likely outcomes. If you are playing a game with two six-sided dice, there are $6 \times 6 = 36$ possible outcomes for each roll. Six of these add up to the number 7, more than any other total possible from rolling of two dice. So the number 7 isn't so much lucky in a game of dice, it's just a sensible assessment of the most likely outcome. This is why you have a good chance of winning at Monopoly if you buy all the properties that are around 7 moves from the Jail square and stack them high with hotels. People tend to land on Jail most often out of all the squares, either by just visiting as a result of their roll or through the various mean cards that send them there (and to get out they have to roll the unlikely combination of a double six). So if they start a turn on the Jail square they are most likely to roll a 7, or thereabouts, and will land in your empire of hotels and houses and pay you a sky-high rent. (A note of caution: One of us has now been playing this strategy for the past two years against an army of ruthless 7–9 year olds and has yet to win a single game.)

Throw total	Possible throws	Number of throws with that total
2	(1, 1)	1
3	(1, 2) (2, 1)	2
4	(1, 3) (2, 2) (3, 1)	3
5	(1, 4) (2, 3) (3, 2) (4, 1)	4
6	(1, 5) (2, 4) (3, 3) (4, 2) (5, 1)	5
7	(1, 6) (2, 5) (3, 4) (4, 3) (5, 2) (6, 1)	6
8	(2, 6) (3, 5) (4, 4) (5, 3) (6, 2)	5
9	(3, 6) (4, 5) (5, 4) (6, 3)	4
10	(4, 6) (5, 5) (6, 4)	3
11	(5, 6) (6, 5)	2
12	(6,6)	1

Meaning from maths

The number 7 also happens to be one of our lucky numbers but we like it because it somehow feels like the first prime number with interesting behaviour. For many people with a mathematical bent, it's the mathematical properties of numbers that make a natural number stand out amongst the rest.

The winner of our highly unscientific poll of *Plus* readers was the number 73. This number seems to have become many people's favourite when the character Sheldon Cooper from the comedy

The Big Bang Theory decreed that it was the best number. His justification came from a confluence of mathematical properties: 73 is the 21st prime number; its mirror, 37, is the 12th prime number; the mirror of 12, 21, is the product of 7 and 3; and 73 in binary, 1001001, is a palindrome, which is the same whether you read it forwards or backwards.

Not coincidentally, Sheldon Cooper shared his reasoning in episode 73, which reminded us of a quote from Martin Gardner (or rather, his imaginary friend, Dr Irving Joshua Matrix, the world's most famous numerologist): 'Every number has endless unusual properties.' (Gardner was a brilliant mathematical popularizer who inspired generations of mathematicians with his 'Mathematical Games' column for *Scientific American*.) Matrix made his statement after coming up with a long list of interesting mathematical properties for the number 2187, part of Gardner's childhood address in Oklahoma. But most interesting of all the mathematical properties of 2187 is that it is a *lucky number*.

Mathematically speaking, a *lucky number* is a survivor of a sieving processes related to the famous sieve of Eratosthenes used to sieve out prime numbers – you will meet it in Chapter 16929639. . .270130176. Like the prime number sieve, the lucky number sieve starts with all natural numbers: 1, 2, 3, . . . As 2 is the first number after 1, you start off by removing every second number:

1, 3, 5, 7, 9, 11, 13, 15, 17, 19, 21, . . .

As 3 is the next number left after the first sieving process, you remove every third number from those remaining:

1, 3, 7, 9, 13, 15, 19, 21, . . .

Then, as 7 is the next number left, you remove every seventh number from those remaining:

1, 3, 7, 9, 13, 15, 21, . . .

And so you carry on, gradually winnowing down the natural numbers to just those lucky survivors.

While the prime number sieve removes numbers based on their prime divisors, the lucky number sieve removes numbers purely based on their position in the number line. So it is particularly surprising that the lucky numbers share many important properties with the primes. There are infinitely many lucky numbers and they become sparser as you move up the number line; the density of lucky numbers (the number of luckies below a certain value) is similar to that of the primes. In fact, the density of twin lucky numbers (those that are 2 apart) is also similar to that of the twin primes. There is even a lucky version of Goldbach's Conjecture (which we met in Chapter 2) – every even number is the sum of two lucky numbers – that is true for at least the first 100,000 integers but, like the prime version of the conjecture, a proof continues to elude mathematicians.

Lucky number 13

The lucky number sieve seems to add weight to our suspicions that the number 7 is lucky. But then, by the same reasoning, so is the number 13.

Poor number 13 has become notorious as an unlucky number, sometimes even being missed out in the numbering of floors in buildings or seat rows in planes. By the mathematical definition, it is in fact a lucky number. Even better, it's a lucky prime. But it seems to have garnered its unlucky reputation from various murky and irrational sources. One theory is that it stems from fact that there are just over 12 lunar months in every solar, hence calendar, year. So every now and again, a 13th full moon has to sneak in which some cultures took to be a portentous event. (The next time a double full moon appears in a calendar month is 31 July 2015, followed by 31 January and 31 March in 2018.)

We can find other arguments from religion: there were thirteen people at the last supper, and rumour has it that Judas was the last to sit down; and when twelve Norse gods had a banquet, Loki, the 13th god, turned up uninvited and all sorts of nastiness ensued. The number 13 has ended up with such a bad rap that the fear of 13 is even a recognized phobia: *triskaidekaphobia*.

Number 13 has some other well known friends that are viewed as unlucky for similarly dubious reasons. The number 4 is unlucky in China as it sounds very similar to the word for death when spoken aloud. And the number 666, widely feared as the 'number of the beast' from the Bible, is thought to come from the Roman emperor Nero, known for his murderous and tyrannical deeds. The Greek spelling of Nero Caesar is *Neron Kaisar*, which, when translated to Hebrew (*nron qsr*) can be converted into a numerical equivalent as the alphabet was also used to write numbers. Thus Nero's name in Hebrew is equivalent to the number 666. If Nero's mum had named him Dave, some other number would have ended up with the bad rap.

Meaning from me

More often than not, we pick our lucky or unlucky numbers from some sort of personal or collective experience. For example, one of us has the favourite number 3, simply because it is the date of her birthday. It's obvious the dates of our or our loved ones' birthdays are very special to us (and a cause of great stress if you forget someone's, say your mum's, birthday, speaking from traumatic personal experience!) but you can be sure they aren't particularly lucky in general. That said, the strategy of picking lottery numbers based on birthdays isn't a bad one: it has exactly the same chance (of approximately 1 in 14 million) of coming up as any other combination you could choose. But the bad news is that you wouldn't be the only person using such a strategy and so you, and they, would be restricting your choices to the same set of numbers, 1 to 31. A combination of birthdays might have the same chance as any of being the winning numbers, but a greater chance of sharing the jackpot if your numbers come up.

In the end, most of the reasons we come up with for preferring some number or another are full of personal and cultural baggage; even if it just means we prefer a number for its mathematical properties, it really is just another subjective point of view. The thing to remember is, just pick the right number for the job. Which is what the next chapter is all about.

10 A matter of scale

We might still think in dozens when we're referring to eggs, but for most other things ten sets the scale. We talk about hundreds of people, thousands of dollars and millions of pounds. It's the powers of 10 that separate one order of magnitude from the next, and that's only natural given that, as we saw in Chapter 0, our number system is based on the number 10.

Scientists and calculators put this idea to good use. Because of the way our number system works, multiplying a whole number by 10 corresponds to appending a zero at the end of the number:

$$4 \times 10 = 40.$$

Therefore, multiplying a whole number by 10, n times, corresponds to appending n zeros to the end of the number. And of course, multiplying by 10 n times is the same as multiplying by 10^n. So an easy way of writing a very large number that has many zeros at the end, such as

4,000,000,000,000,000,000,000,000,000,000,000,000, 000,000,

is to write it as a product involving powers of 10, in this case

$$4 \times 10^{45}.$$

And this is what scientists and calculators usually do. In *scientific notation* large numbers are represented in the form

$$a \times 10^b,$$

where the number a is at least 1 but less than 10 and b is the appropriate whole number. This also works when the number isn't made up of just a single digit followed by lots of zeros. For example, the mass of the Earth,

5,972,200,000,000,000,000,000,000 kg,

can be conveniently written as

5.9722×10^{24} kg.

Adding 1 to the exponent (power) of 10 tells you how many digits there are in the number (in the case of the mass of the Earth, 5.9722×10^{24} kg, adding 1 to the exponent of 24 gives 25 digits) and the number you multiply the power of 10 by (59722, with the decimal point removed) gives you the first few digits of your number.

What if you are dealing with a very small number? As we saw in Chapter *e*, a number raised to a negative power, for example 10^{-2}, is equal to 1 divided by the number raised to the same power but positive, for example

$$10^{-2} = \frac{1}{10^2} = 0.01,$$

so you can write very small numbers in terms of negative powers of 10. For example, the mass of an electron is about

0.000 000 000 000 000 000 000 000 000 000 910 938 221 kg,

which in scientific notation works out as $9.10938221 \times 10^{-31}$ kg. This time the exponent minus 1 gives you the number of zeros after the decimal point (30 in this example) and the number you multiply by (91093822, with the decimal point removed) gives you the digits that follow the zeros.

And this reveals the beauty of scientific notation. The mass of the Earth is 5.9722×10^{24} kg. The mass of the electron is $9.10938221 \times 10^{-31}$ kg. By comparing exponents you immediately see that the Earth is 55 orders of magnitude heavier than an electron. That's because to get from something on the scale of 10^{-31} to something on the scale of 10^{24}, you need to multiply by 10^{55} (see Chapter e for rules of powers). Using scientific notation you get an immediate feel for the size of things and how they compare, without the drudgery of counting digits.

Very large

The majority of numbers we meet in everyday life (such as sizes, distances, weights and amounts) move in a limited range, probably somewhere between −100 and 100. That's not a fact of nature but, at least in part, a consequence of the units we choose to measure things by. A metre is the length it is because it compares well to our own size and the scale at which we live our lives.

WHAT'S THE SCALE?

Getting an idea of scale is often easier if we can think in terms of things that we are familiar with: for example, describing something as being the length of 3 London buses is easier to visualize than 25.14 m. Here are some useful comparisons:

Width of human hair: 10^{-4} m

Length of a grain of rice: 5×10^{-3} m

Height of a human: 1.7m

Length of a London bus: 8.38 m

Eiffel Tower: 324 m

Mount Everest: 8,848 m

Great Wall of China: 6,400 km

River Nile or Amazon: 6,600 km

Earth: diameter 12,756 km, distance round equator 40,075 km, mass 5.9×10^{24} kg

Distance to moon: 380,000 km . . . and back again 760,000 km

Grand piano: 240–450 kg

Elephant: 3000–7000 kg

Blue whale: length 33 m, mass 1.8×10^5 kg

Head of a pin: 10^{-6} m² (about 1 square mm)

Postage stamp: 5×10^{-4} m²

Football pitch (UK): 100–110 m long, 7,140 m² in area

Wales: 20,779 m²

Olympic-size swimming pool: 2,500 m³

In science, however, things can become very, very large. Galactic distances are so vast that they are measured not in kilometres or miles, but in light years: the distance light can travel in a year. Since light moves at roughly 300,000 kilometres per second, and there are

$$60 \times 60 \times 24 \times 365 = 31{,}536{,}000$$

seconds in a year, a light year corresponds to roughly

$$300{,}000 \times 31{,}536{,}000 = 9{,}460{,}800{,}000{,}000 \text{ km.}$$

That's around 9.4608×10^{12} km in scientific notation. It's a humbling fact that Andromeda, the nearest significant galaxy to Earth, is 2.5 million light years away. That's

$$(9.4608 \times 10^{12}) \times (2.5 \times 10^{6}) \text{ km} = 23.652 \times 10^{18} \text{ km} = 2.3652 \times 10^{19} \text{ km.}$$

In the grand scheme of things we really are very, very small!

And very small

In physics things can become unimaginably small but still be of huge significance. In 1905 Albert Einstein published a paper to explain the *photoelectric effect*: if you shine a beam of light onto a metal surface then the light can knock out electrons from that surface. This surprising phenomenon had puzzled physicists since the late 1880s. If you think of light as a wave, which was customary at the time, then you would expect the energy with which an electron flies off to be the same as that of the wave that knocked it out. The energy of the light wave is proportional to its intensity, so the energy of the ejected electrons should depend on the intensity of the light that is shone at them.

But experiments showed that it didn't. Instead, the electron's energy seemed to depend on the *frequency* of the light. This was weird: why should the frequency of the wave, the number of times it peaks in a given time interval, have anything to do with the electrons' energy? The only thing that did depend on the intensity of the light was the number of electrons that were knocked out: the brighter the light, the more electrons were broken loose.

Einstein explained the curious effect by suggesting that in some situations light could also be thought of as a stream of particles, which are called *photons*. The intensity of the beam of light is proportional to the number of photons in it, so the more intense the beam, the more electrons get knocked out by individual photons. Each photon comes with a precise amount (or *quantum*) of energy *E*, which is imparted to the knocked-out electrons. It is this amount *E* that is related to the frequency *f* of the corresponding light wave. The relationship, as recognized by Einstein, is

$$E = hf,$$

where *h* is an incredibly small number. It's 6.626069×10^{-34} (when measured in units of m^2 kg/s). This number *h* is called *Planck's constant* after the physicist Max Planck who first came across it.

Written in ordinary decimal notation this gives a number with a zero before the decimal point and 33 zeros after it.

Waves matter

Nearly twenty years after Einstein's explanation of the photoelectric effect, which earned him the Nobel Prize in Physics in 1921, a young French physicist by the name of Louis de Broglie boldly suggested that not only light but also matter can be thought of both in terms of waves and in terms of particles. It's a strange idea, but since de Broglie's suggestion it has been confirmed in many experiments: particles, such as electrons, behave like waves in some aspects and like ordinary particles in others. The relationship between the frequency of the 'particle wave' and the energy of the particle is exactly $E = hf$.

This *wave–particle duality* became the central tenet of quantum mechanics, that highly counterintuitive theory that describes the world at very small scales. Planck's constant h describes the relationship between wave and particle aspects, so it's not surprising that it plays a central role in that theory.

Planck seems to have been well aware of its significance even before the theory of quantum mechanics was fully fledged. In 1899 he proposed units to measure length, mass, time, temperature and electric charge that were made up of h and some of the other accepted constants of nature. Planck thought that his units were the real deal, utterly independent of human experience. As he wrote in his paper,

> *These [units] necessarily retain their meaning for all*
> *times and for all civilizations, even extraterrestrial and*
> *non-human ones, and can therefore be designated as*
> *'natural units'...*

The reason why Planck's units haven't caught on in everyday life is that they are very, very much smaller than the kind of quantities we come across. A metre corresponds to roughly 10^{35} Planck length units, that's a 1 followed by 35 zeros – just imagine your average IKEA flatpack instruction book with all lengths given in such astronomical numbers!

The end of known physics

In physics, however, the Planck length plays a special role, which becomes apparent when we try to combine Einstein's theory of general relativity with quantum mechanics. At first sight these two theories stay well out of each other's territory. General relativity describes the behaviour of large objects such as planets and stars, while quantum physics describes tiny particles. But size isn't the only thing that matters; mass does too.

If you compress an object of a given mass to a small enough region of space, then its gravitational pull will become so strong that nothing, not even light, can escape from its vicinity. You have created a *black hole*. To turn the Earth into a black hole you'd have to compress it to a radius of 9 mm, about the size of a peanut. The length scale at which a given mass turns into a black hole has a name: it's called its *Schwarzschild radius*. To describe a black hole we need general relativity, so we can take the Schwarzschild radius as measuring the length scale at which general relativity becomes crucial in describing an object of a given mass.

But what is the length scale at which quantum physics kicks in? Again, this depends on mass. In addition to its Schwarzschild radius, an object of a given mass m also comes with another number, called its *Compton wavelength*, which you can take as measuring the length scale at which quantum physics becomes crucial in describing the object. The question is, is there a mass m for which the two length scales are equal, so that to describe an object of that mass and size you need both theories simultaneously?

An approximate back-of-the-envelope calculation suggests that there is, and that it is exactly the Planck mass. The corresponding length scale is the Planck length. So to talk about an object of Planck mass and length you need a theory that encompasses both gravity and quantum physics – and, as we have seen in Chapter 4, such a theory does not yet exist. Experimental physics doesn't encounter this problem because the Planck length is far smaller than anything we could observe. It is about 10^{-20} times the size of a proton (while the Planck mass is about the mass of a large cell). But still, the Planck length and mass seem to mark the boundary at which our current theories fail.

This argument is naive in many ways, but it gives a clue as to why the Planck units are likely to play a special role in a theory of quantum gravity when it is eventually found. In many versions of string theory, which is an attempt at quantum gravity (see

Chapter 4), the tiny oscillating strings that represent particles are assumed to be of roughly the Planck length. And some people even suggest that the Planck length is the smallest length possible. If you zoom in on space to that level, you will see it breaking up into pixels of Planck size: there is nothing smaller than the Planck length and space itself cannot be subdivided into smaller pieces. With current techniques it is utterly impossible to zoom in on space to that level to see what is really going on – so we may have to wait some time to see whether all this is really true. Talking about time, we should devote some of it to the number that embodies it: 12.

12 About time

No one really knows why there are 12 hours on the clock. The habit goes back to the ancient Egyptians, who divided the sunlit part of the day into 10 hours, plus an hour each for dusk and sunset. The number 12 may have come about because there are roughly 12 lunar cycles in a year, and if you divide the year into 12, then why not divide the day in the same way? Some people say it was because the Egyptian number system had 12 as a base, and others because Egyptians were using the number of joints in their fingers to count. Each finger has three joints; if you exclude the thumb because that's the digit you use to count the others, then you have $4 \times 3 = 12$ joints on each hand.

Whatever the reason, the Egyptian approach meant that the length of an hour varied throughout the year, since the sunlit part of the day is shorter in winter than it is in summer. It seems that the Greeks were the first to ignore whether it was actually light or not and decide to divide the day-and-night into 24 equal hours.

Today we not only ignore daylight and darkness, we even ignore the poor old Earth itself. Modern, standardized and coordinated time has nothing to do with the Earth's rotation; instead it is set by the resonance frequency of the caesium atom measured by a few hundred carefully guarded atomic clocks dotted around the world. The speed at which the Earth rotates is variable and unpredictable, whereas the atomic oscillations are completely regular. In fact, *coordinated universal time* (Temps Universel Coordonné, UTC) and Earth time would slowly drift apart if left alone, so every now and again a leap second is added to the UTC. It gets inserted between 23:59 and 0:00 of either 30 June or 31 December, and is shown on UTC clocks as 23:59:60. The impressively named International Earth Rotation and Reference Systems Service (IERS) announces a few months before each date whether a leap second will be added, based on what the Earth has been doing.

This preoccupation with exactitude seems more than a little obsessive. Perhaps it's because our inbuilt sense of time isn't very accurate: it seems to stretch or contract depending on whether we are in a boring meeting or passionately kissing. Today we are thoroughly enslaved by time and consider it absolute. Even if the Earth were to stop in its tracks you'd still have to keep your appointments. As Newton put it, 'absolute, true and mathematical time, of itself, and from its own nature, flows equably without relation to anything external'.

Time for symmetry

It's a good thing that time is measured in cycles. Had the first people to start measuring the hours decided to count beyond 12, or even 24, and gone on to say that it's 25 o'clock, 26 o'clock, 27 o'clock and so on, then the time right now would be a number so large it would take a considerable amount of time to even say it.

In calculations, the cycle of time can cause a few problems at first. On a 12-hour clock, the numbers 1 to 12 form their own little world which you cannot leave by adding or subtracting hours. For example, on a 12-hour clock 7 + 7 is no longer 14, it's 2. Similarly, 3 – 7 is no longer –4, it's 8. Each number in this 12-hour world has a sibling, which when added to it brings you back round to 12. For example, the sibling of 7 is 5 and the sibling of 2 is 10. Subtracting a number, x, of hours is the same as adding its sibling, $12 - x$. What's the sibling of 12? Well, it's 12 itself, since 12 + 12 = 24, which in the 12-hour world is the same as 12. In fact, adding 12 to anything leaves it unchanged, so 12 is to the 12-hour clock what 0 is to ordinary arithmetic. This is why 12 o'clock midnight is also known as 0 o'clock.

This mildly interesting feature of the clock is an example of one of the most important concepts of mathematics: that of a *group*. Mathematicians define a group abstractly as a collection or set of objects and a way of combining two of them to get a third (an *operation*). In our example the set of objects is the numbers 1 to 12, and the way of combining two of them to get a third is addition *modulo* 12. The set of objects qualifies as a group if it meets four rules:

1. When you combine two objects from the set the result is also a member of the set.
2. There is an *identity object* in the set, which when combined with any other object, leaves it unchanged (the identity is 12 in our example).
3. Every object comes with a sibling, called its *inverse*, so that when you combine the two the result is the identity.

4. The combining operation doesn't care about bracketing: it's *associative*. Writing + for this operation and *a*, *b* and *c* for three objects from the set we have $(a + b) + c = a + (b + c)$. Clearly this works for our clock example, for example

$(1 + 2) + 3 = 3 + 3 = 6$ and $1 + (2 + 3) = 1 + 5 = 6$;

$5 + (1 + 8) = 5 + 9 = 2$ and $(5 + 1) + 8 = 6 + 8 = 2$.

Same but different

The nice thing about groups is that two of them can consist of different objects but still have exactly the same structure. The group for the 12-hour clock is cyclic: you can generate the whole thing by repeatedly adding 1s, and once you get to 12 you start from the beginning again. Now think of a standard circular clock face which has the 12 hours marked, but no numbers written next to them. A clockwise rotation around its centre through a twelfth of a full turn (30 degrees) moves one hour dash to the next one along, so it leaves the clock face unchanged: it's a symmetry. Applying this symmetry once corresponds to adding 1 to an hour on the clock face. Applying it twice (so the total angle of rotation is 60 degrees, or two twelfths of a circle) corresponds to adding two hours. And so on. The rotations generated in this way are the ones through angles that are multiples of $\frac{1}{12}$, from $\frac{1}{12}$ itself through to a full turn. When you have done a full turn you start at the beginning again, so the collection of clockwise rotations through an angle that's a multiple of $\frac{1}{12}$ form a cyclic group in exactly the same way as our numbers 1 to 12 with addition modulo 12.

The two groups are what is called *isomorphic*: although one consists of numbers and the other of rotations, they have exactly the same structure.

The collection of *all* the symmetries of the clock face, not just the rotations but also the various reflections that are possible, form a bigger group, which contains the rotational symmetries as a subgroup. In fact, the collection of symmetries of any shape form a group and this is why group theory is often described as the mathematical language of symmetries. But although people have been fascinated with symmetry for millennia, the abstract study of groups only came into its own in the 19th century. It was inspired by objects whose symmetries are far from obvious: equations.

Able Abel

Ever since Babylonian times people have been studying the solutions to equations, such as

$$x^2 = 9.$$

The solutions to this equations are 3 and –3; if you substitute these numbers for x the result is 9, as required by the equation. Another example of a *quadratic* equation ('quadratic' because the highest power of x is 2) is

$$x^2 - x - 1 = 0,$$

which has solutions $\frac{(1 + \sqrt{5})}{2}$ and $\frac{(1 - \sqrt{5})}{2}$. The first of these is the golden ratio, which we met in Chapter ϕ.

There are infinitely many different quadratic equations, so there's no way of solving them all one by one. But luckily, we can represent a *general* quadratic equation by writing

$$ax^2 + bx + c = 0,$$

where a, b and c are understood to stand for some given numbers (for example, $a = 1$, $b = -1$, $c = -1$ in the equation that gave the golden ratio). Even more luckily there is a general formula that gives us the solutions:

$$\frac{(-b + \sqrt{b^2 - 4ac})}{2a} \text{ and } \frac{(-b - \sqrt{b^2 - 4ac})}{2a}.$$

You can try this out by plugging $a = 1$, $b = -1$, $c = -1$ into this general formula to see that you do indeed get the two solutions for the golden ratio equation.

Forms of this general solution were already known to the Babylonians around 4000 years ago, but it wasn't until the sixteenth century that mathematicians worked out general solutions for equations in which the highest power of x is 3 or 4:

$$ax^3 + bx^2 + cx + d = 0$$

and

$$ax^4 + bx^3 + cx^2 + dx + e = 0,$$

where the numbers a, b, c and d are rational numbers (i.e. whole numbers or fractions: see Chapter $\sqrt{2}$).

After that considerable achievement (you will find out more in Chapter i), the hunt was on for a general solution of the *quintic equation*, in which the highest power of x is 5, but by 1800 it still hadn't been found. In 1824 the Norwegian mathematician Niels Henrik Abel delivered a bombshell which explained why: he proved that there isn't one. The quintic equation can't be solved by a general formula that involves only adding, subtracting, multiplying, dividing and taking roots. This doesn't mean that a quintic equation can never be solved – you could try and do it by clever guessing techniques or if you were lucky the equation might be nice enough to make a solution possible. For example,

$$ax^5 + b = 0.$$

This has one solution, namely the fifth root of $\frac{-b}{a}$.

But the point is that you can't write down a general solution for the general version of the quintic. Abel delivered this result, which the mathematical elite of Europe had worked on for at least 250 years, at the tender age of 22. He died five years later from tuberculosis and in abject poverty. Although the great mathematicians of the day had recognized his genius, he had never managed to get a job.

Symmetry and tragedy

Not long afterwards, another mathematical prodigy, Evariste Galois from France, decided to get a better grip on why the general quintic should be unsolvable. He did this in the midst of a busy schedule of political bravado which, following the unrest of 1830, got him arrested twice. Once for apparently threatening the life of King Louis Phillipe (not directly but by proposing a toast at a dinner) and once for wearing the uniform of the Artillery of the National Guard, which was illegal, while carrying loaded guns and a dagger. But Galois had good reason to be confident. Since the age of sixteen he had been sure of his considerable mathematical talent and it was only poor performance at routine exams and unruly behaviour that had hampered his young academic career.

Galois started playing around with solutions of equations and noted that in a particular mathematical sense you could swap them around, just as you can swap around the corners of a triangle or other regular shape by applying one of its symmetries. You can get a glimpse of these equation symmetries if you look at the plus-and-minus reflection found in the solutions of the general quadratic equation earlier. Whether or not an equation is solvable depends on how the solutions can be swapped around – it depends on the underlying symmetries. Galois developed a theory of symmetry for equations which explained why there isn't a solution for a general quintic: in general the symmetries don't behave in the right way. It was while he was thinking about this problem that Galois developed the foundations of group theory.

And it wasn't any time too soon. In 1832, while evacuated from Paris during a cholera epidemic, Galois seems to have fallen in love with a woman called Stephanie-Felice du Motel. For unclear reasons that were probably connected to Stephanie he was challenged to a duel by Perscheux d'Herbinville, set to take place on 30 May 1832. The night before the fateful day Galois jotted down his mathematical discoveries with instructions to a friend on where to publish them should he not survive. The next morning he met his adversary and died from his wounds the following day. He was only 20 years old.

The biggest theorem ever?

Galois' work eventually led to what's arguably the greatest mathematical project ever undertaken. Since two groups can have the same structure even if they are made up of different things (as in the clock example earlier) mathematicians prefer to treat them as abstract collections of 'things', called *elements*, without specifying what those things are. You can simply represent the elements by letters and describe the group by writing down how all the different elements, denoted by different letters, combine (see the following box for an example). The question now is this: what kind of structures can you find in a group? The group that came from the clock is cyclic and its period of repetition is 12. But if you look at other groups, say the group of symmetries of a wallpaper pattern as explored in Chapter 5, the structure might be different. In general a group can contain smaller subgroups that relate in various ways to the big group that they live in. Is it possible to classify all groups that are made up of finitely many elements?

*	1	a	b	ab
1	1	a	b	ab
a	a	1	ab	b
b	b	ab	1	a
ab	ab	b	a	1

This table represents the so-called Klein four group. It consists of four elements: 1 (the identity), a, b and ab. The table shows how these combine. For example, a combined with ab gives you b. You see this by finding a in the column on the left, ab in the row at the top, and then seeing what's in the cell corresponding to them both.

The answer is yes, but not without considerable effort. The so-called *classification of finite simple groups* was completed in 2004 and ran to tens of thousands of pages, arranged in hundreds of journal articles, involving over 100 different authors from around the world. Finite simple groups are the building blocks that other groups are made up of. There are infinitely many of them, but the classification tells us that each belongs to one of 18 families that are well understood, or is one of 26 groups that don't fit in anywhere else and are called *sporadic groups*. The largest sporadic group is aptly called *The Monster* and consists of

808,017,424,794,512,875,886,459,904,961,710,757,005,754,368, 000,000,000

elements!

The completion of the classification had first been announced in the 1980s, but it turned out to contain a flaw which took some years to fix. Michael Aschbacher, one of the authors of the proof, expressed a surprising degree of uncertainty, for a mathematician, when he wrote in 2004, 'to my knowledge the main theorem [of our paper] closes the last gap in the original proof, so (for the moment) the classification theorem can be regarded as a theorem'. There are probably fewer than five people in the world who understand the proof of this enormous result in its entirety. And even now at least three research groups are working on reformulating it. The theorem may be considered complete but the project to understand and explain it continues. Now, from the biggest mathematical proof to something rather smaller . . .

ε It is a truth universally acknowledged . . .

. . . that ε (pronounced 'epsilon') is always a very small number, and usually comes with a δ (pronounced 'delta').

42 Mostly harmless

The 2013 Nobel Prize in Chemistry was awarded to three scientists who helped take chemistry out of the lab and into cyberspace. Martin Karplus, Michael Levitt and Arieh Warshel developed ways of simulating complex chemical systems on a computer. They found the optimal balance between competing theories to describe those systems: classical physics going back to Newton, which has equations that are easy to solve but which does not describe the world accurately at very small scales; and quantum physics, which does, but involves a huge computational effort.

An initial inspiration came from a substance called *retinal* which, as the name suggests, exists in the retina of the eye. Today chemists are using the tools the Nobel laureates developed for all sorts of purposes, from making environmentally friendly cars to producing medical drugs. But Michael Levitt, so it was reported, has a far more ambitious goal: he would like to simulate whole living organisms at the level of molecules.

If Douglas Adams is to be believed then such undertakings don't have to be as Frankenstein-esque as they first appear. In his cult classic *The Hitchhiker's Guide to the Galaxy* the whole Earth is a giant supercomputer built by an alien race from the planet Magrathea to find the question of life, the Universe and everything. They already know the answer from a previous supercomputer and it's 42 – leaving them no choice but to search for the actual question to make sense of it. In the Hitchhiker's guide, the indispensable companion for intergalactic travellers, the Earth is described by two words: 'mostly harmless'.

Some things in Douglas Adams' 'trilogy in five parts' have almost come to pass since the story was first conceived in 1978. The Hitchhiker's guide described in the book is a small-ish electronic device containing information which, if written down, would fill up several planets. In the 1970s this must have seemed far-fetched and fanciful, but now that smartphones are putting the internet at our fingertips wherever we are, it seems quite plausible.

What we haven't quite managed – in fact we are very far away from it – is to simulate systems as complex as a whole planet with everything that lives on it. We can't even simulate the weather accurately enough to predict what will happen next week, and when it comes to the economy we seem to be stumped too. A research paper published in 2013 claimed that some of the predictions of the Reserve Bank of Australia were no more accurate than what you'd get if you based your predictions on rolling dice.

Climate, it's rocket science

So what's the problem? Let's avoid all things human made (except, perhaps, carbon emissions) and try to simulate the Earth's climate, in order to predict it. Since climate is about general trends, such as the average annual temperature in some large region of the Earth, this should be easier than the weather, which is about details like whether it will rain in a particular part of Haringey, North London, on Wednesday. The first thing to do is to divide the surface of the Earth up into a grid and do the same to vertical space, cutting it into chunks extending from the bottom of the oceans to high up into the atmosphere. The result is a collection of grid boxes. The climate in each of these boxes today will affect the climate in each of its neighbours tomorrow, as winds and atmospheric circulation move things around.

The next step is to capture the way each box impinges on its neighbours by mathematical equations. This may seem like a daunting task, but luckily we have known the key equations for quite some time. Descriptions of winds and ocean currents are based on the *Navier–Stokes equations*, which have been around since the middle of the 19th century, developed with major input from the French engineer and physicist Claude-Louis Navier and the Irish-born mathematician and physicist George Gabriel Stokes. Given the starting conditions of a fluid flow, the equations will (in theory) give you its speed, direction of motion and pressure at every point in space and future time. To describe how the temperature of the oceans and the atmosphere varies, we can use thermodynamics, a theory which was also fully fledged by the end of the 19th century.

In climate models equations like these are linked to capture as much of the physics as possible. Now simply feed in the starting values for each grid box and get a computer to calculate what the climate will do tomorrow, the day after, and so on, all the way into the future. This, in a nutshell, is how climate models work.

Those butterflies again

The idea is neat, but producing sophisticated global climate models is much more complicated than it looks. For starters, the equations involved are incredibly difficult to solve. The Navier–Stokes equations describe anything from large-scale movements of the gulf stream right down to the tiny fluctuations that come from someone speaking and creating turbulence in the air. The amount of complexity they capture is so vast, it's not surprising the answers don't just pop out. And it's not just that solutions are hard to find. No one knows if physically meaningful solutions exist for the most general form of the equations. There is even a price on their head – whoever eventually finds the solutions, or proves that they don't exist, is set to win $1 million from the Clay Mathematics Institute.

This latter problem is one for the pure mathematicians, as in practice scientists make do with approximate solutions. But even these require an immense amount of computing power. For a model with a horizontal grid spacing of 1 km to simulate climate for the next century would require a computing power on an enormous scale of 10^{18} operations per second. Such power isn't currently available, so scientists are working with a much coarser grid. In typical global climate models the horizontal grid spacing is between 200 km and 600 km and the vertical spacing a few kilometres.

The problem now is that we run into the famous butterfly effect (see Chapter $\sqrt{2}$), the idea that the tiny air disturbance created by the flap of a butterfly's wing in Brazil can cause a tornado in Texas. It's this effect that makes it impossible to forecast the weather for more than a few days in the future as weather can hinge on tiny random fluctuations. This isn't to say that climate forecasting is immune to the butterfly effect, far from it. For example, thunderstorm systems in the tropics, which are only a few hundred metres or a few kilometres in size, can influence climate variability on the scale of thousands or tens of thousands of kilometres.

This means that you can't ignore what happens within the grid boxes. One way of dealing with this is to represent things that happen there in a simplified way. For example, you might assume that the amount of clouds in a grid box depends on the relative humidity in the air in a much simpler way than it does in reality. Another, surprising, method is to add a little randomness into your model. Rather than representing variables in a simplified way you decide their values at random, metaphorically speaking by casting a die. You need to prepare that die carefully though, making sure that the likelihood of different outcomes reflects what you see in real life. It sounds paradoxical, but it's been shown that these kind of *stochastic models* can outperform their traditional counterparts in terms of accuracy.

A word for skeptics

There is no denying that climate modelling, let alone weather forecasting, comes with uncertainty – you may not have captured all climate phenomena in your model, limited computing power means a limit to the resolution, and there's the butterfly effect to contend with. Even if lots of extra computing power suddenly became available, the butterfly effect wouldn't go away. Edward Lorenz, who coined the term (see Chapter $\sqrt{2}$), demonstrated that climate phenomena at different scales, from small turbulences within clouds all the way up to large jet streams, can interact in such a way that improving the precision of initial observations beyond a certain level has only a negligible effect on the accuracy of prediction.

Since the predictions issued by climate scientists are usually pretty alarming, climate change sceptics take pleasure in pointing out those uncertainties. What they fail to mention, though, is that a large part of climate science is about quantifying and understanding this uncertainty. Scientists at the Intergovernmental Panel on Climate Change (IPCC) use a sophisticated combination of different models, and careful analysis of their predictions, based on four different scenarios of how the concentration of greenhouse gases in the atmosphere might vary in the future in response to our behaviour. They are also very careful to phrase their predictions using probabilities instead of certainties.

And the predictions are still worrying. According to the IPCC's Fifth Assessment Report, for all but the most optimistic scenario (for which greenhouse gas emissions drop to zero by about 2070) there is at least a 66% chance that the global mean surface temperature will have increased by more than 1.5° by 2100, compared with the period 1850–1900. The two higher-concentration scenarios say there's a greater than 66% chance it will have increased by more than 2%. The latter figure, 2%, is recognized as a critical level of increase beyond which things get seriously dangerous, but even 1.5% could spell disaster in some areas of the world. 66% is by no means certain – but if you think that it isn't certain enough, then ask yourself if you would board a plane that has a 66% chance of crashing.

Building brains

If climate is hard to simulate accurately, then what about that most indispensable part of human life, the human brain? It's made up of nerve cells called neurons connected by links called synapses. A complete wiring diagram of this network would be every neuroscientist's dream, but unfortunately it's far too large. According to estimates there are around 85 billion neurons in the human brain, wired up by over 100 trillion interconnections. What is more, connections can shift and change, making a complete model of the human brain the stuff of sci-fi.

But that's no reason to give up. After all, the brain isn't just one big mess of neurons, it can be sub-divided into different parts fulfilling different functions, and neurons come in many different species too. So even if we can't model a brain in all its details, at least we might hope to understand its components and its large scale structure.

One fascinating fact about the brain is that it seems to exhibit similar connectivity as networks in other areas of life and nature. In 2011 the Cambridge psychiatrist Ed Bullmore demonstrated this beautifully in a lecture at the Cambridge Science Festival. He asked Twitter users to tweet during the lecture using a particular hashtag. After the lecture he displayed the image showing the connectivity of the hashtagged tweets. The network, which had evolved naturally without any conscious design effort, displayed a characteristic we already met in Chapter 6: it was a small-world network. What is more, it was made up of chunks, each containing nodes that are densely connected to each other, but sparsely connected to nodes in other chunks.

Studies have shown that the human brain exhibits the same two features: they are small-world networks and they are made up of individual modules that are sparsely connected to each other. This makes evolutionary sense. Small-worldness means that the average distance between nodes in the network (measured in the number of hops to get from one to the other via other nodes) is relatively small. This means that information can get around the whole network quite easily, so information transfer is globally efficient. But small-worldness also means that there is a high degree of local clustering – neighbours of neighbours tend to be neighbours of each other too. This means that there are many ways for information to get around local neighbourhoods, so there is also local efficiency. The modularity of the brain, the fact that it's made up of chunks, makes sense because that way modules can be changed one at a time without threatening the function of others – a highly important feature for a system that needs to adapt, evolve and change.

Information transfer is obviously hugely important, but brains face another problem that Twitter networks avoid: they are metabolically expensive. The human brain is about 2% of body mass, but it consumes about 20% of the body's energy budget. A lot of the metabolic money goes into sending signals down synapses – the links between neurons – which means that brains face a similar trade-off as circuits in computer chips. On the one hand they need to be complex to ensure high performance, but on the other they need to minimize wiring cost, the sum of the length of all the connections.

Intriguingly, designers of computer circuits and nature, the designer of brains, seem to have evolved the same approach to solving this trade-off. Studies conducted by Bullmore and colleagues suggest that both networks follow a surprising mathematical relationship, first discovered in computer circuits by the IBM employee E.F. Rent in the 1960s. It's connected to their modular nature. Suppose that by cutting through a few connections you've partitioned your network into chunks, each chunk consisting of around N nodes. Then the number C of connections that link each chunk to the rest of the network (that is, the number of connections you've had to snip through to cut that chunk loose) is roughly equal to

$$C = kN^p.$$

The numbers k and p don't depend on your individual chunk, or on the number N, but are characteristic of the network as a whole. The number k is the average number of connections per node in your system as a whole. The number p, called the *Rent exponent*, is between 0 and 1. For high-performance computer circuits as well as human brains p is typically in the region of 0.75.

What's interesting is that computer chip designers didn't explicitly set out to build circuits to follow this rule. Rather, it seems to have evolved because networks need to be simple enough to be buildable. In a totally random network, one that doesn't exhibit any structure in the way the nodes are connected up, you might expect the number C (the number of connections from a chunk of the network to the rest of the network) to grow in direct proportion to N (the number of nodes in the chunk): the more nodes there are in a chunk, the more connections there are. So you might expect a relationship of the form

$$C = kN.$$

In other words, a random network has a Rent exponent $p = 1$.

However, neither nature nor computer chip designers would build a network in such a random fashion. It's much better to build it up module by module, placing and connecting things cleverly to minimize wire length as much as possible. The reduced Rent exponent seems to be a result of a hierarchical design process, which favours short-range over long-range communication.

So it seems that nature and computer scientists have come up against similar problems and solved them in similar ways. In mechanics there's something called the *principle of least action*: all physical systems, from the orbiting planets to the apple that supposedly fell on Newton's head, behave in a way that minimizes the effort required. You can use the principle to derive the fundamental laws of motion. Some people have suggested that Rent's rule represents a similar fundamental law of information processing, a result of a principle of least cost. But we'll need a lot more research into the nature of information to find out if this is really true.

Do we live in a matrix?

The fact that we are currently unable to simulate the brain, the climate and definitely not the whole planet doesn't rule out a shocking possibility: that we ourselves live in a computer simulation run by humans in the far-distant future or perhaps even an alien race. In fact, if you assume that our technological expertise is likely to keep growing and that our descendents will be as interested as we are in simulating nature, then this possibility seems increasingly likely. Unless, that is, you think that humans will die out before they reach that highly advanced state. Arguments along those lines have led philosophers to take the outlandish-sounding *simulation hypothesis* more seriously than most of us would find comfortable.

But how can we ever find out? One approach is to assume that our future hyper-intelligent descendants would base their simulations of the world on the same techniques that physicists use today to simulate the interactions of the fundamental particles of nature. With a better understanding of the theory and unheard of amounts of computing power, our descendants might then be able to simulate not just tiny regions of space but the whole Universe, or perhaps even several of them. Current techniques work by dividing spacetime up into a grid, just as climate models divide up the Earth and its atmosphere. If we live in a simulation that works along the same lines, then this grid may be detectable. In 2012 the physicists Silas R. Beane, Zohreh Davoudi and Martin J. Savage proposed ways in which one could test this idea. If we live in such a pixelated artificial Universe then high-energy cosmic rays should travel through space in a certain way that gives away the underlying grid, and not interact equally in all directions as we would expect if there were no grid. If observations show that rays don't behave in the grid fashion, then there is some reason to believe that we don't live in a simulation that works along the lines described above. If they do, then the possibility can't be ruled out.

But, whatever the behaviour of cosmic rays turns out to be, it won't count as conclusive proof. There is still the possibility that, just as in *The Hitchhiker's Guide to the Galaxy,* we live in a simulation run by an alien race with entirely different technological and scientific means. In the *Guide* the Earth is demolished just before completing its purpose – to find the question to the ultimate answer to life, the Universe, and everything – because it had to make way for a hyperspace bypass. Its two surviving inhabitants went on to experience some mind-blowing adventures, including eating dinner at the restaurant at the end of the Universe. This was probably far too posh to serve the kind of food stuffs we will think about next.

43 Would you like fries with that?

When you go to a well-known fast-food outlet to buy golden brown pieces of something that calls itself chicken you are faced with choices. Traditionally those nuggets were sold in boxes of 6, 9 and 20. This meant that if you wanted to eat precisely 15, 33 or perhaps 38 nuggets you were in luck. All these numbers can be made as a sum of 6s, 9s and 20s, and were therefore called McNugget numbers. If, however, you wanted to eat precisely, say, 11, 16 or 43 nuggets you would have been disappointed, as no combination of boxes would have given you those. This is why they were called non-McNugget numbers.

The number 43 had the special honour of being the largest non-McNugget number. Any whole number you could think of, as long as it was bigger than 43, could be purchased using the appropriate combination of boxes (perhaps with some box sizes repeated). Ever since the well-known fast-food outlet introduced boxes of four nuggets, the largest non-McNugget number has been 11, but in the minds of (some) mathematicians 43 still retains its special status as the one and only original largest non-McNugget number.

Similar questions to the nugget one come up in other situations too. If you're a cashier who has run out of everything but 2p, 10p and 20p coins you might wonder what amounts of change you can make up with them. In Rugby Union the question is what point scores are possible given that individual goals score either 3, 5 or 7 points. In both cases it's clear that there are some small numbers you can't make up as a sum of the numbers on your given list, for example 1p in the coin example or 2 points in rugby. But is it always true, as in the nugget case, that once you have a large enough number, you can be *sure* that you can make it out of your given components? Is there always a largest non-McNugget number for a given list of numbers?

The answer is yes, but only if the numbers in your given list are just right. In the coin example, all the numbers in the list – 2, 10 and 20 – are even. Any sum made out of even numbers will also be even, so you cannot make odd numbers. Since there isn't a largest odd number, there then isn't a largest non-McNugget number. The same applies if the numbers on your list are all multiples of 3, 4, 5, or any other whole number k greater than 1. In that case any sum would also be a multiple of k, and since there isn't a largest multiple of a number k, there isn't a largest non-McNugget number in this case either.

If, however, the numbers on your list are not all multiples of some other number k greater than 1, if they are *relatively prime* as is the case in the Rugby Union example, then there always is a largest non-McNugget number. That's a fact that can be proved mathematically. In Rugby Union the largest non-McNugget number is 4, which means that every point score greater than 4 is possible in theory.

Although expressing numbers as sums of other numbers seems like a simple problem, as often happens in number theory, such a simple problem can quickly become tricky. In the nugget example, a formula giving the largest non-McNugget number for a given list of relatively prime whole numbers only exists if the list contains at most two numbers. If these two numbers are n_1 and n_2, then the corresponding largest McNugget number is

$$n_1 n_2 - n_1 - n_2$$

(unless one of n_1 or n_2 is equal to 1, in which case you can clearly make up all positive whole numbers, and the largest non-McNugget number is 0). Once you have more than three numbers in the list, however, there is no known formulaic solution and you need to find the largest non-McNugget number using computer algorithms.

Let me count the ways . . .

We could of course turn the question around. Instead of asking what numbers we can make by adding those in a given list of positive whole numbers, we can ask in how many ways we can write a particular number as a sum of any positive whole numbers. For example, the number $n = 5$ can be written as:

5

4 + 1

3 + 2

3 + 1 + 1

2 + 2 + 1

2 + 1 + 1 + 1

1 + 1 + 1 + 1 + 1.

So there are seven ways of writing 5 as a sum of positive whole numbers. Each way is known as a *partition* of 5. It's clear that, as the number n grows, the number of partitions (called the *partition number*, $p(n)$) grows too, and grows incredibly quickly. For example, the number of partitions of the first few powers of 10 are

$$p(1) = 1,$$
$$p(10) = 42,$$
$$p(100) = 190,569,292,$$
$$p(1,000) = 24,061,467,864,032,622,473,692,149,727,991.$$

To put that last number in perspective, the number of partitions of 1,000 is greater than 2.4×10^{31}, a number with 32 digits. The number of partitions for the next power of 10, $p(10,000)$, is over 3.6×10^{106}. The largest number for which the number of partitions has been computed, using sophisticated computational techniques, is 10^{19}: the answer, $p(10^{19})$, is a huge number with just over 3.5 billion digits. It may have a simple definition, but the number of ways to partition a number quickly gets out of our control. An obvious question is: can we tame these unexpectedly wild numbers?

Partitions and pentagons

This problem attracted the attention of Euler, the prolific Swiss mathematician we have already met many times in this book. Euler discovered an ingenious way of calculating the number $p(n)$ of partitions of a whole number n, based on the number of ways to partition numbers smaller than n.

If you step through this process from the start it doesn't seem very logical. You start with $p(0) = 1$. Then

$$p(1) = p(0) = 1,$$
$$p(2) = p(1) + p(0) = 2,$$
$$p(3) = p(2) + p(1) = 3,$$
$$p(4) = p(3) + p(2) = 5.$$

So far, so like the Fibonacci series. But next we have

$$p(5) = p(4) + p(3) - p(0) = 7,$$
$$p(6) = p(5) + p(4) - p(1) = 11,$$
$$p(7) = p(6) + p(5) - p(2) - p(0) = 15.$$

What is going on here? In fact, underlying this seemingly random choice of adding and subtracting previous values of the partition function is an ingenious recursive formula based on the *pentagonal numbers*. These are the number of dots needed to draw successively larger nested pentagons, like this:

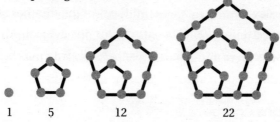

1   5   12   22

The first four pentagonal numbers: 1, 5, 12, 22

Instead of drawing the pentagons, you can also generate these numbers with a formula. This isn't as strange as it might at first appear: we all learnt how to generate *square numbers* in primary school with the formula n^2. The formula for pentagonal numbers is

$$\frac{(3n^2 - n)}{2},$$

which gives the first four pentagonal numbers shown in the picture above (with 1, 5, 12 and 22 dots) for $n = 1, 2, 3, 4$. If we let n be any whole number, positive or negative, then we get what is called the *generalized pentagonal numbers*:

$$1, 2, 5, 7, 12, 15, 22, 26, \ldots$$

by taking n to be 1, –1, 2, –2, 3, –3, 4, –4, and so on.

Euler used these generalized pentagonal numbers to define his surprisingly simple recursive formula for the partition number of any number n:

$$p(n) = p(n-1) + p(n-2) - p(n-5) - p(n-7) + p(n-12)$$
$$+ p(n-15) - \ldots$$

This recursive formula gives the exact value of the partition numbers. It is incredibly useful for calculating the first few partition numbers, but it isn't feasible for calculating the number of partitions of an arbitrarily large number – the computational effort involved soon becomes huge.

Partition patterns

After Euler, partition numbers continued to intrigue mathematicians, including amateur ones: everyone knows how to add numbers so there is plenty of scope for playing around with the problem, and perhaps even making a new discovery. At the beginning of the 20th century the problem attracted the attention of one amateur mathematician whose ability turned out to tower over that of many of his professional contemporaries.

Srinivasa Aiyangar Ramanujan was born in 1887 in a small village about 400 km south-west of Madras (now Chennai) in India. His mathematical abilities became apparent quite early on. At the age of 15 he had worked out how to solve quartic equations all by himself and he worked on mathematical problems that went way beyond what his classmates were dealing with. But since he put little effort into subjects other than mathematics, he failed to get into university and continued his studies alone and in poverty. News of his genius spread through Indian mathematical circles though, and a plea for help to one eminent mathematician, who described Ramanujan as 'A short uncouth figure, stout, unshaven, not over clean', eventually got him a job as a clerk in the accounts section of the Madras Port Trust.

It was during his time as a clerk that he wrote a letter which was to change his life. It was addressed to the well-known Cambridge number theorist Hardy (whom we met in Chapter 2) and contained a number of mathematical results. Some were original, some were re-discovered, many were difficult and some, according to Hardy, appeared to be 'new and important'. Ramanujan provided no proofs and it was clear he had no formal training, but Hardy was so intrigued by this apparently self-taught genius that he invited Ramanujan to Cambridge. In 1914 Ramanujan set sail for England to begin one of the most remarkable collaborations in mathematical history.

Among the many things Ramanujan was interested in were the partitions of positive whole numbers. The ultimate goal to understanding partitions would, of course, be a neater formula giving the partition number, $p(n)$, of a positive whole number n in a single line of mathematics. Ramanujan did not quite get there, but he and Hardy found a formula that gave the approximate answer. A little later on, in 1937, Hans Rademacher perfected their techniques and managed to give an exact formula for the number of ways a number could be partitioned. Though exact, this formula was incredibly complicated, involving infinite sums and still requiring huge amounts of computing power to work out the number of partitions for any large numbers. In practice, only an approximate version of the formula could be used (with the infinite sums truncated and the answers rounded off) leading to approximations.

We had to wait more than 70 years, but mathematicians Ken Ono (US) and Jan Bruinier (Germany) finally discovered a finite formula for the partition numbers in 2011. And in a pleasing twist, the machinery from number theory that they used – called *mock modular forms* and *Maas forms* – originated from mysterious mathematics found in the very last letter Ramanujan wrote to Hardy just months before he died.

Ramanujan made many other important contributions to number theory, including work on the Riemann zeta functions which play a central role in the famed Riemann Hypothesis (see Chapter 2). Unfortunately his life came to an end too soon. He had suffered health problems from a young age and during his stay in Cambridge these worsened due to the unaccustomed food and cold weather. Ramanujan returned to India in 1919 and died the following year at the age of 32. He is remembered as one of India's greatest mathematicians.

60 An adventure through time and space

The next time someone asks you what use mathematics is, point to your watch or point to a map. The division of an hour, or a degree of latitude or longitude, into 60 minutes, and a minute into 60 seconds, is a direct consequence of the Babylonians' choice of 60 as the base of their place value number system (see Chapter 0). Although we don't know why the Babylonians chose 60, it proved to be a particularly useful choice for calculating fractions and ratios as 60 has so many factors. In fact 60 is the smallest number to have 12 factors (which are 1, 2, 3, 4, 5, 6, 10, 12, 15, 20, 30, 60) and no other number under 100 has more factors. It is a *highly composite number*: it has more factors than any smaller number, a concept first defined and studied by none other than Ramanujan (whom we met in the last chapter) in 1915.

This is particularly handy when you compare it with our more familiar decimal system, with base 10, which has only four factors (1, 2, 5, 10). Any fraction in our decimal system has either a finite decimal expansion, such as $\frac{1}{5} = 0.2$, or an infinite one that ends in a repeating block, such as $\frac{2}{9} = 0.2222\ldots$ The same is true in number systems that use other bases, such as the sexigesimal system (base 60). Any fraction $\frac{a}{b}$ (in its simplest form, so we can't cancel out any common factors from a and b) will either have a finite expansion or an infinite one ending in a repeating block of digits.

The former only happens if the denominator, b, has the same factors as the base of the number system. Therefore, in the decimal system only fractions whose denominators are products of 2s and 5s have a finite expansion, while in the sexigesimal system there is more leeway. This means that many rational numbers (see Chapter $\sqrt{2}$) with an infinite decimal expansion have a finite sexigesimal expansion (which we write with a semicolon ';' acting in the place of a decimal point, and commas separating the sexigesimal places).

Already in these first few fractions, the infinite decimal expansions of $\frac{1}{3}$, $\frac{1}{6}$ and $\frac{1}{9}$ can be written as finite sexigesimal expansions.

Fraction	Decimal	Sexigesimal
$\frac{1}{2}$	0.5	0;30
$\frac{1}{3}$	0.333...	0;20
$\frac{1}{4}$	0.25	0;15
$\frac{1}{5}$	0.2	0;12
$\frac{1}{6}$	0.1666...	0;10
$\frac{1}{7}$	0.142857 142857...	0;8,34,17,8, 34,17...
$\frac{1}{8}$	0.125	0;7,30
$\frac{1}{9}$	0.111...	0;6,40
$\frac{1}{10}$	0.1	0;6

This number system allowed Babylonian mathematicians incredible accuracy, for example they approximated the value of $\sqrt{2}$ to three sexagesimal places

$$\sqrt{2} \approx 1 + \frac{24}{60} + \frac{51}{60^2} + \frac{10}{60^3},$$

which is within 8 millionths of the real value. It was after more than three millennia and the advent of the Renaissance that any culture could improve on their accuracy. This led many ancient mathematicians to use the Babylonian number system rather than their inferior Greek numerals (which were not even a place value number system, as we saw in Chapter 0). In the second century BC, Ptolemy wanted greater accuracy in his astronomical calculations and observations, and so following the Babylonians' lead he subdivided each of the 360 degrees in a circle into 60 *partes minutae primae* (which we now know of as *minutes*) and each of these into 60 *partes minutae secondae* (now known as *seconds*).

Time and space

Although Einstein is famous for bringing together space and time into spacetime in his 1905 theory of special relativity, space and time have been inextricably linked throughout human history. Our very first understanding of the passage of time was through the movement of celestial objects (the Sun and the Moon) across the sky.

If you stargaze for any length of time (or take one of those stunning time-lapse photos), many stars seem to trace out circles in the night sky. So it isn't surprising that humanity first thought that the stars were fixed to a giant celestial sphere that rotated about the Earth or, a little later, one that the Earth rotated within. One of the ways of measuring the passing of time at night was by the movements of these stars, and the timings of the seasons were predicted based on the positions of the Sun and the Moon against the background constellations (the zodiac) on the celestial sphere. So astronomy, the mathematical study and prediction of the movement of stars, was vital to understanding the passage of time.

Astronomy was also vital in navigating on Earth. Latitude and longitude as we know them were first defined by the ancient Greek astronomer and mathematician Hipparchus, who lived in the second century BC in Greece and Egypt. As well as making many astronomical observations of his own, he also used the star catalogues of the Babylonian astronomers, which were produced from 1200 BC to Hipparchus' time. From these pooled observations Hipparchus was able to calculate the length of the year with impressive accuracy, to within just 6.5 minutes of its true value.

Latitude from the stars

The only fixed points in the night sky are those that lie in line with the axis of rotation of the Earth (or, equivalently, the axis of the celestial sphere). The North Star, Polaris, lies almost directly above the North Pole. There is no equally obvious South Pole star. To find the *south celestial pole* you need to intersect a line running through the long axis of the Southern Cross (a constellation of stars) and another line that bisects the pair of stars called the Southern Pointers.

Hipparchus calculated the latitude at a location on Earth to be the angle that Polaris (or the south celestial pole in the southern hemisphere) sits above the horizon. If you are at the North Pole, Polaris will be directly overhead at an angle of 90 degrees to the horizon. At the equator it will lie almost exactly on the horizon. In fact at any location in the northern hemisphere, the angle that Polaris sits above the horizon is exactly the angular distance of the location from the equator thanks to some simple trigonometry.

To see this we drop down to two dimensions and examine the circle that we get from slicing the Earth cleanly through the North Pole and our location at the surface, P. Polaris lies right above the North Pole. But it is so far away that we can take the line of sight to Polaris from any point on the Northern Hemisphere to

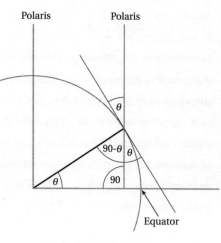

Calculating latitude

be parallel to the vertical direction from the North Pole to Polaris. If we look North towards the horizon, our line of sight just touches the Earth at our location P and makes an angle, which we'll call θ, with the vertical line to Polaris.

The radius of the Earth at the point P meets the line of sight to the horizon at a right angle. Therefore, the angle between the radius at P and the line to Polaris must be (90 − θ)°. This angle forms part of a right-angled triangle, where the final side runs from the centre of the Earth towards the equator (horizontal in our diagram), meeting the vertical line through Polaris at right angles. Then the *angular distance* from the equator, otherwise known as the latitude of P, is the angle between the line to the equator and the radius at P. Given that the other angle in the right-angled triangle is $(90 - \theta)°$, the angular distance from the equator to P is just our angle θ, the angle Polaris sits above the horizon.

Time for longitude

As for longitude, Hipparchus made the entirely sensible suggestion that places that experienced the same local time as given by the Sun should be on the same line of longitude. But without a reliable timekeeper this was not an easy definition to use in the real world. Indeed, calculating accurate longitude eluded astronomers and navigators alike for nearly two thousand years. It cost countless lives lost at sea and vast sums of money. A famous example is the Scilly naval disaster of 1707 in which four ships were lost and over 1,400 sailors died just off the Isles of Scilly. Bad weather didn't help, but a major cause of the disaster was the sailors' inability to figure out exactly where they were.

Many seafaring countries, including Spain in the 16th century, the Netherlands in the 17th century and Britain in the 18th century, offered huge sums of money to those who could solve the longitude problem. In the end, the problem was not solved by one of the many stars of science and mathematics who tackled it, such as Galileo, Cassini, Hooke, Newton and Wren. Instead the answer came from a British watchmaker, John Harrison. Harrison developed a truly ingenious clock (which resembles an oversized pocket watch) that did not rely on a pendulum and so could keep accurate time at sea.

Once you have an accurate way to keep time, calculating longitude is a relative doddle. The Earth rotates a full turn (360 degrees) in 24 hours, so in each hour it would rotate by roughly 15 degrees. One of the prizes of winning the longitude wars was laying claim to the *prime meridian*, or 0 degrees longitude, which is taken to run through Greenwich, London. If you set your clock to Greenwich Mean Time (the time at Greenwich as defined by the Sun), then at noon in Greenwich the Sun is at its zenith (the highest point it reaches in the sky) for the day. However, at the same time in a different longitude, say 15 degrees east, such as somewhere in Germany, the Sun has already passed its highest point in the sky and is 15 degrees closer to the horizon. At noon Greenwich Mean Time, the angle the Sun has moved from (if you are east of Greenwich), or has yet to move to (if you are west of Greenwich), its zenith at your location allows you to calculate your longitude. (The calculation is a little more complicated but this gives you the general principle.)

Where am I?

Using the stars and a clock you can pinpoint where on the globe you are, but that information would be fairly useless without that other great navigational tool, the map. This too poses mathematical challenges. The problem is that the Earth is round, while a map is flat. The only way to represent the surface of a sphere on a flat sheet of paper is to introduce distortion. You can convince yourself of this by wrestling an orange out of its skin, keeping the skin in a single piece. When you put the skin down, it will naturally retain its rounded shape; any attempt to flatten it will either break it or stretch and squeeze it in some direction.

Therefore, any flat map of the Earth is inaccurate in some respect. The most familiar one is the Mercator projection, named after the 16th-century Flemish cartographer Gerardus Mercator. The basic idea behind this projection is to put the Earth (or, better, a shrunk version of the Earth) into a 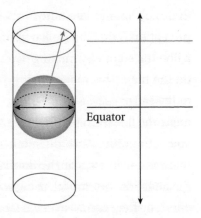 vertical cylinder, touching at the equator and with the North Pole pointing straight up. Given any point x on the Earth you then draw the line that connects x to the centre of the Earth. Extend that line until it hits the cylinder at some point y – that point y is the *projection* of your point x. Using that technique you can trace the outlines of continents and countries onto the cylinder, which you can then cut open and unroll into a flat sheet.

A problem arises, however, as you near the poles. The projected points move higher and higher up the cylinder as you converge on the North Pole and further and further down the cylinder as you approach the South Pole. To project the whole Earth you'd need a cylinder that is infinitely long. The two poles themselves don't appear on the cylinder at all: the line that connects them to the centre of the Earth is parallel to the sides of the cylinder and therefore never meets it.

Mercator's projection is a variation of this cylindrical projection, which ensures that angles are represented faithfully on the map. The problem with the poles still applies, however, and it's for this reason that the Mercator projection doesn't contain them and their immediate surroundings. The great advantage of this projection is that lines of constant compass bearing (for example, a line traced out by moving due east) correspond to straight lines on the map. This means that if you want to sail from A to B, you only need to draw a straight line between the two on the map. The angle the line makes with the equator (or any line of latitude) tells you your north–south direction and its relation to the Greenwich meridian tells you your east–west direction. You then head off in the direction thus found, using your compass as a guide. As long as you keep the compass needle fixed and don't hit any islands along the way, you will get there.

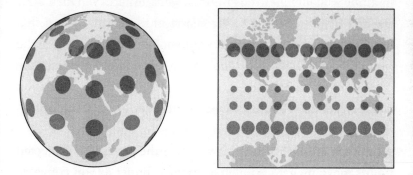

The Mercator projection. The grey circles correspond to circles of equal area on the globe.

The disadvantage of the Mercator projection is that, as a direct result of the cylinder technique, distances near the poles are stretched, making places near the poles appear much larger than they actually are. On the standard projection Greenland appears to be larger than Africa, when in reality Africa is over 14 times bigger! Politically, this has awkward consequences. Europe and North America happen to sit far enough North of the equator to benefit from the distortion. They appear far bigger in relation to the belt surrounding the equator than they are in reality, thus adding cartographical discrimination to cultural and economical discrimination.

There are other projections which attempt to redress the balance. The Gall–Peters projection represents the relative sizes of areas faithfully – and you'll be surprised how small Europe really is – but it greatly distorts shapes. Near the equator, shapes are squeezed horizontally and stretched vertically, with the opposite happening near the poles. Continents that cross the equator appear long and thin, which is why the projection has been compared to a washing line on which countries have been hung out to dry.

You are here

Today we hardly ever rely on paper maps any more. Our ability to navigate has been revolutionized by our smart phones, with their maps and ability to know exactly where we are. The GPS (the Global Positioning System that we met earlier, in Chapter 4) feature of our phones was originally developed by the US Department of Defence and only available to the military. However, in response to serious navigation disasters, such as a civilian aircraft being shot down when it strayed into Soviet airspace in 1983, the GPS became available to everyone just a few years later. (This US government policy was a modern echo of the UK parliamentary Longitude Act established in response to naval disasters over three centuries earlier.) At first the GPS equipment was both cumbersome and expensive, but in recent years it has reduced in size and price so that now many of us have a GPS device in our cars, with satnav, or even in our pockets as a feature of our phone or camera.

The system relies on a network of 31 GPS satellites that circle the globe, with at least four overhead at any point in the world at any time. We can precisely predict the orbits of these satellites and know with great accuracy their location at any time. The satellites constantly broadcast messages, giving the time and the satellite's location when the message was sent. GPS receivers, like the one in your phone, listen out for these messages, calculating the time it took the message to travel from the satellite to you. Then, using the familiar equation

$$distance\ travelled = speed \times time,$$

and the fact that the message was a radio signal travelling at the speed of light (approximately 300,000 kilometres per second), your phone can calculate your distance from the satellite. So if the message took 0.06 seconds to reach you, you are $0.06 \times 300,000 = 18,000$ kilometres from the satellite. From this information your phone knows that you are somewhere on a sphere with a radius of 18,000 kilometres, centred on the location of the satellite when it sent the message.

To narrow down our location further, your phone listens for the message from two more satellites (remember that there are at least four overhead at any moment) and performs a similar calculation for each. You now know that you are on three different spheres, centred on each of the three satellites. Generally, two spheres intersect in a circle. The third sphere will intersect that circle in two points. Since only one of these two points is on Earth, three satellites are enough to pinpoint your location.

So the state-of-the-art technology of GPS is as simple as that – it amounts to nothing more than intersecting spheres and using a basic equation relating speed, distance and time. If you pinpoint your location using transmitters on Earth, rather than satellites, then the problem becomes even easier. It reduces to two dimensions and you only need to intersect circles, rather than spheres.

In the real world there are, admittedly, a few further minor complications. As we know, three satellites are enough to locate our position. So why are there always at least four satellites overhead at any one time? The problem is that the timings involved have to be incredibly accurate. Atomic clocks, that gain or lose less than a microsecond in a century, are used in the satellites but these are way too expensive and ungainly to fit in every phone and satnav. Instead, while your location can be pinpointed with three exact measurements, it can also be found with four or more less precise measurements.

Where are you?

Circles are one of the *conic sections* – curves produced by slicing through a double cone with a flat plane. You will have seen these yourself if you've ever played with a torch on a dark night. If you shine the torch directly up at the ceiling, you produce a circle of light that grows or shrinks in size depending on how far away you

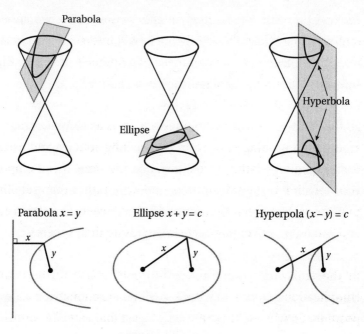

Parabola

Ellipse

Hyperbola

Parabola $x = y$

Ellipse $x + y = c$

Hyperpola $(x - y) = c$

The conic sections

hold the torch. The ceiling is acting as a flat plane slicing the cone of light from the torch and it produces a circle of light if the ceiling is perpendicular to the axis of the cone. If you angle the torch a little, the circle distorts to an ellipse. Angle the torch a little more and the light will suddenly streak off across the ceiling to infinity (if you had a very big room), creating a shape called a *hyperbola*. And the shape you get at the point at which the light just tips from an ellipse to a hyperbola (when you are angling the torch so that the side of the cone of light is parallel to the ceiling) is called a *parabola*.

Each of this family of curves has a geometric description that gives it fascinating mathematical properties. A circle is the set of all points that lie at the same distance from the centre (the focus of the circle). This can be thought of as a special case of an ellipse, which has two foci rather than just one. An ellipse is the set of all points for which the sum of the distances to the two foci is always the same ($x + y$ equals a constant c in the image on page 221). A hyperbola also has two foci, but this time it's the difference of the distances from the two foci to a point on the hyperbola that is always the same ($|x - y| = c$, where the vertical bars indicate the absolute difference, turning negative into positive where necessary). The parabola is defined by one focus and a line called the *directrix* (the line in the picture at the bottom of page 221): a point lies on the parabola if its distance to the focus is the same as its distance to the directrix.

The conic sections have fascinated mathematicians for thousands of years. The Greek mathematicians Euclid and Archimedes were studying them as far back as 300 BC. Archimedes used a particularly useful property of parabolas to design a heat ray that was intended to set fire to an enemy ship. Rays (of light, sound or other energy) that enter a parabola running parallel to its axis of symmetry will be reflected off the curve and meet at the focus. Thus, the combined heat or sound of all these rays is concentrated at the focus. Archimedes proposed arranging many mirrors in a parabolic shape that would capture the heat of the Sun's rays and concentrate them on the focus of the parabola. If a ship (one of your enemies, it was hoped) was at the focus of this large parabolic arrangement of mirrors, it would feel the full force of the captured heat of the Sun and, with any luck, catch fire.

This property of the parabola also has more modern applications in the shape of satellite and telescope dishes. These parabolic dishes capture weak incoming rays (whether light and radio waves from distant stars or the transmission of your favourite show via satellite TV) and focus them on a receiver at the focus of the dish.

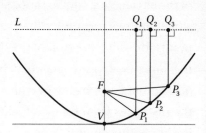

Rays concentrated at the focus of the parabola

The hyperbola too has a very modern application, not dissimilar to the function of GPS. Suppose you want to locate the position of a target that is sending out signals using a transmitter. You pick up their signal (perhaps a radio transmission, a distress beacon or just a cry for help) in two receivers at different locations. This signal has travelled an unknown distance x to the first receiver and an unknown distance y to the second receiver. If these distances are the same ($x = y$) then the signal will arrive at the two receivers simultaneously. It is more likely, however, that the target is closer to one receiver than the other, resulting in a slight time difference between when the signal arrived at each of the two receivers. Multiplying this time difference by the speed of the signal gives the difference in the distances from the target to the two receivers: $|x - y| = c$. This means the target

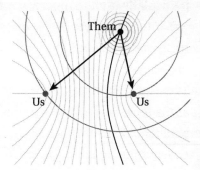

The difference in arrival time of the signal to two receivers allows us to locate the target on a hyperbola with foci at the two receiver locations.

lies somewhere on the hyperbola satisfying $|x - y| = c$ with foci at each of the two receivers.

If you can pick up your target's signal at three receivers you can use this process to locate the target on a different hyperbola for each pair of receivers. This would allow you to pinpoint the target's location at the intersection of the three hyperbolae, just by listening out for a signal they have sent. They don't even have to know you are listening, or indeed want to be found by you, which is what makes this technique, called *multilateration*, particularly powerful. As well as being used today in air traffic control and search and rescue missions, it has also been used for covert surveillance, such as in the Second World War to locate enemy artillery ranges by the sound of their gunfire.

So you can try to hide, but maths is probably going to find you. Do you want to take that chance?

100% You can be (almost) certain of this

Looking at the grey skies outside my window this morning, I'm finding it hard to believe the forecast of only 1% chance of rain today. That seems overly optimistic for almost any day in the UK, let alone an overcast day in September. Despite the technical and mathematical sophistication of weather forecasting, we all have stories of when forecasts have been wrong. It's a good example of how unexpected events do happen and usually when you've forgotten your umbrella.

We deal with probability, risk and chance every day of our lives. We might choose to eat or avoid certain foods because of research showing they will increase or decrease our chances of developing diseases. The most prominent example is smoking, which within our lifetime has gone from being viewed as a harmless activity to one responsible for increasing your risk of lung cancer. But, again, there are always stories of someone who smoked every day of their adult lives and lived to be 103.

How do we actually go about measuring chance? One way goes back to the 17th century, when a gambler's dispute led people such as Blaise Pascal and Pierre de Fermat (of the famous Fermat's last theorem; see Chapter $\sqrt{2}$) to develop a mathematics of chance. The idea is to consider the symmetries of your set-up. Take, for example, a coin. It has two sides and, unless it is somehow wonky or irregular, each side has an equal chance of coming up when you flip the coin. This means that you have a 50% chance (or a probability of $\frac{1}{2}$) of flipping heads. The same reasoning gives you the chance of winning the lottery. There are around 14 million possible outcomes (13,983,816 to be precise) because that's how many different six-number combinations you can draw from 49. Assuming that all outcomes are equally likely, that is, that the lottery machine doesn't favour any of the balls, this gives you a winning chance of around 1 in 14 million.

In many real-life situations you can't use this symmetry approach, so you turn to proportions instead. If your doctor tells you that you have a 5% chance of getting cancer, then that's because studies have shown that out of a large population of people similar to you, 5% did get cancer. Forecasts involving a percentage chance of rain often come from *ensemble forecasting*. As we saw in Chapter 42, weather is very sensitive to small changes in initial conditions. Given that we can't perfectly measure at every point on the globe the temperature, pressure and other variables that feed into weather forecasts, meteorologists instead run many simulations of their weather models, each with slightly varied initial conditions that are similar to real observations. Then if in 1% of these ensembles of forecasts it rains today, they forecast a 1% chance of rain today.

Behind this approach lies the *frequentist* interpretation of probability: the idea that the probability of an event (such as a coin coming up heads) should be interpreted as the proportion (also called *relative frequency*) with which that event would happen in a large number of trials (coin flips). This is at loggerheads with another interpretation, called *Bayesian* after the 18th-century mathematician and Presbyterian minister Thomas Bayes, which views probabilities as subjective quantities, which measure how confident we are that an event will occur, based on available evidence. If new evidence comes to light, we can then update our numbers.

The (f)law of averages

However you choose to interpret probability or measure it, there is a whole area of mathematics devoted to it. Its rules are relatively straightforward, but its outcomes don't always chime with intuition. We're all familiar with the law of averages, also known as the law of large numbers: when you flip a coin a large number of times, and heads and tails have an even chance of coming up, then you'd expect the share of heads and tails in your sequence to be roughly 50:50. Therefore, if you have just seen a coin flipped 999 times and only heads came up, you'd be tempted to bet on tails at the next flip, as surely this imbalance must be redressed.

But you'd be wrong, and you'd join millions of gamblers in their misfortune. The next coin flip is independent of all the others, so your chance of tails is only 50%, as it was for every one of the previous flips and will be for all the future ones. This seems in conflict with the law of large numbers, but it isn't. What the law tells you is that, as you increase the number of coin flips towards infinity, the proportion of heads will converge on 50%. Even if you have flipped a very large number of heads, there's still an infinity of flips left to bring the average down to 50%, so the single event of your next flip is under no obligation to come up tails to conform to the law.

There are plenty more examples of the slippery nature of chance. Many books have been written about it, some for entertainment, and some for serious philosophical reasons (more on this below). What is for sure, however, is that our struggles with chance are intimately linked to our struggles with its twin sibling: randomness.

It's normal to be random

Without the aid of some random process, like flipping a coin, we humans are very bad at generating randomness. If I ask you to write down a random sequence of heads and tails, of numbers, or to draw dots at random across the page, you won't be able to help yourself from taking the previous choices into account when making the next one. It's often quite easy to spot that such sequences haven't been generated randomly because we are so careful to ensure a nice even spread of our choices. Our brains are naturally geared to pick up on patterns and this is why we find it so hard to generate randomness. As soon as we notice the first hint of any pattern we think we've broken the randomness and alter our next choice. No matter what Destiny's Child said, we are not good at being independent when it come to making random choices.

In fact, in a truly random sequence of events, any combination of events is possible including all the nice looking patterns. A string of 10 heads in a row, HHHHHHHHHH, is just as likely as

HTHTTTHHTT, TTHHHTHTHH, or TTTTTTTTTH. Any sequence of 10 coin flips is as likely as any other.

Despite over a millennium of experimental research into randomness (otherwise known as gambling), filled with flipping coins and rolling dice, the concept proved surprisingly slippery to pin down mathematically. The first person to formally tackle the concept of randomness was the French mathematician Émile Borel. In addition to being a brilliant mathematician who helped found an area of maths called *measure theory*, he was also politically active, serving as Minister of the Navy and fighting in the Resistance during the Second World War.

In 1909 Borel considered numbers with infinite decimal expansions (or an infinite expansion in any other base; see Chapter 60). Borel defined such a number as *normal* if all digits appear in the expansions with the same frequency, all pairs of digits with the same frequency, all triples of digits with the same frequency, and so on. So the number 1 should appear $\frac{1}{10}$ of the time, as should any of the other digits. That's because there are ten digits in all and spreading the frequency evenly gives you $\frac{1}{10}$. By the same reasoning each possible pairs of digits, such as '11', '23' or '09', should appear $\frac{1}{100}$ of the time, as there are 100 of those pairs. The same goes for any triple (appearing $\frac{1}{1,000}$ of the time). A sequence of n digits should appear $\frac{1}{10^n}$ of the time. A normal number does not favour any particular finite combination of digits. By the same reasoning, every finite combination of digits should appear in the number's infinite expansion. If each digit in an infinite sequence was produced by the roll of a fair ten-sided die, then normality is exactly what you'd expect to get. That's why normality can be interpreted as a condition for randomness in an infinite sequence: if it's to be random, it has to be normal.

Borel showed that normal numbers exist and that 'most' real numbers are normal (he had a clear definition of what he meant by 'most'). Mathematicians believe that all our favourite mathematical constants – including π, e and $\sqrt{2}$ – are normal, but so far no one has been able to prove that any of them are. Borel's proof that almost every real number was normal wasn't constructive and he was left with a powerful result but no explicit example.

To the rescue, a few years later, came the lowly undergraduate student D.G. Champernowne. In 1933 he had the blindingly simple yet ingenious idea of stringing together all the natural numbers one after the other, after a decimal point:

> 0.12345678910111213141516171819202122232425262728293031323334353637383940 41. . .

Champernowne's number, as it has come to be known, is lovely not just for its definition but also because it was one of the first examples of a normal number in base 10. Each digit, each pair, each triple and so on of digits appears with exactly the same frequency as the others. And every finite sequence you can imagine appears somewhere within the infinite depths of Champernowne's number.

But just stop for a minute and consider what that means. Champernowne's number, or indeed any other normal number (including more famous suspected examples, such as π which we are very confident is normal), holds every finite sequence you can think of. Every number. Your name, if you translate it into numbers. In fact, it contains the name of every person who has ever lived and ever will live. Every sentence that has been written and every sentence that will be written. Every book that has ever been written and ever will be. The history of the world and every possible future.

Library of Babel

This might first appear to be verging on the mystical or miraculous (if you've seen the movie *Pi*, you'll know that hunting for meaning in such numbers can become an unhealthy obsession) but you need to treat this information in exactly the same way as you would the chance you'll be dealt the perfect poker hand or roll that elusive snake eyes (double one in a pair of dice). Just because something can happen doesn't mean it will happen to you in your lifetime. Just because the full text of this book is already written in the digits of Champernowne's number or (probably) π, doesn't mean we'll ever find it. The first words of Chapter 0 in this book – 'In the' (which is 010010111000000101000100000101 when translated into the binary code computers use) – appears at the 2,385,438,088th digit in the binary expansion of π. But we'll have to look much further for even the next letter (Chapter 0 starts with 'In the beginning . . . '). The binary string for 'In the b' (01001011100000010100010000001010000 000010) does not appear in the first 4 billion binary digits of π. Who knows how many digits we'd have to search to find the first sentence.

This concept was beautifully captured by the Argentine writer Jorge Luis Borges in his 1939 essay *The Total Library* and his 1941 short story *The Library of Babel*. In his story, Borges imagined a universe as a seemingly endless complex of connected identical hexagonal rooms. The shelved walls of these rooms contain books filled with every possible combination of letters and punctuation. As Borges noted in his essay, such a library will indeed hold everything:

> *Everything would be in its blind volumes. Everything: the detailed history of the future, Aeschylus'* The Egyptians, *the exact number of times that the waters of the Ganges have reflected the flight of a falcon, the secret and true nature of Rome, the encyclopedia Novalis would have constructed, my dreams and half-dreams at dawn on August 14, 1934, the proof of Pierre Fermat's theorem, the unwritten chapters of* Edwin Drood . . .

The narrator of the short story explains that the occupants of the library were overjoyed when they first realized that their universe contained all possible books. However, this was soon replaced by a deep depression. The occupants of this library spent a great deal of their lives journeying through the endless identical rooms on a quest to find meaning among the books. But, just as we are yet to find any hidden meaning within the digits of π, the presence of every possible combination of symbols means that virtually all the books in the library will be nonsensical. 'The certainty that some bookshelf in some hexagon contained precious books, yet that those precious books were forever out of reach, was almost unbearable.' The story is a fascinating exploration of the ideas of randomness and normality, and a warning for those who search for meaning (or a winning lottery ticket) in the output of a random process.

Infinite monkey business

We have Borel, he of the normal numbers, to thank for another literary example involving randomness. When he was thinking about statistical mechanics – the bulk behaviour of the zillions of atoms and molecules that make up a gas or liquid – he used the metaphor of a million monkeys typing on typewriters. Even if they typed for ten hours a day for a year, he wrote, it would be unlikely that their combined writings would exactly equal the content of the world's richest libraries. This has come to be known as the *Infinite monkey theorem*, referring to the fact that if an army of monkeys typed randomly for long enough, they would eventually produce the complete works of Shakespeare. But how long would we actually have to wait?

David Spiegelhalter, our favourite statistician at the University of Cambridge, made the calculations when he was helping the BBC to produce a *Horizon* programme about infinity in 2009. Let's imagine there is only one monkey who types one of 31 characters – the 26 letters of the alphabet, a space, a comma, a full stop, a semicolon and a hyphen – whenever it hits the keyboard. We'll ignore any other bits of punctuation and also assume that there are only lower case letters in Shakespeare. The chance that the first character is correct is 1 in 31. The chance that the first two are correct is 1 in 31×31, which works out at 1 in 961. And so on. Since there are around 5 million characters in the complete works, the chance the monkey produces them in one go of typing is 1 in $31^{5,000,000}$.

We can rewrite this as a power of 10 (see Chapter 10) to get a chance of 1 in $10^{7,500,000}$. It's a very small chance indeed, roughly equivalent to flipping a fair coin 25 million times and getting heads every single time, or winning around 1 million lottery draws in a row. So we can be virtually certain that the monkey won't produce the complete works of Shakespeare at its first go at typing 5 million characters, but what if we keep waiting? Spiegelhalter calculated that if the monkey produced 50 characters a second (which is very fast) then to be 99% sure to get even a sequence of only 17 characters right (for example, 'To be or not to b'), we would have to wait for 13.2 billion years – that's about the time that has passed since the Big Bang!

Spiegelhalter's calculations were borne out by the monkey simulator, a computer program that the BBC commissioned to simulate a monkey typing. After 113 million monkey seconds (generating 50 characters a second) the longest match was 'we lover', which occurs in the speech by Boyet, 'With that which we lovers entitle affected', in *Love's Labours Lost*, Act 2 Scene 1.

Real monkeys fared even worse. When an arts project in Paignton Zoo put a computer in a monkeys' enclosure, the animals typed five pages, mainly consisting of the letter 's'. They then lost interest and used the keyboard as a toilet. 'That's a rather limited output of classic English literature,' wrote Spiegelhalter, 'which just shows the problems that turn up when maths meets the real world.'

Pinning down chance

Probability is one of the areas of maths that meets reality most frequently, so we should finally pay some attention to its philosophical foundations. Probability theory gives us rules for calculating with chance. For example, it says that the probabilities of all the possible outcomes of an event should add to 1. This makes sense: if there is a probability of $\frac{1}{2}$ of tossing heads with a coin, then the chance of tails must be $1 - \frac{1}{2} = \frac{1}{2}$. It also tells us how to work out probabilities of combinations of outcomes. For example, if the probability of an outcome A is p and the probability of an outcome B is q, and the two events are independent, then the probability of both A and B happening is $p \times q$. In the coin example, it means that if you flip two fair coins, or the same fair coin twice, then the probability of getting two heads is $\frac{1}{2} \times \frac{1}{2} = \frac{1}{4}$.

The problem is that we can't measure probability as easily as we can measure, say, length or weight. We can use abstract considerations, like saying that a coin flip having two possible outcomes means that the probability of each is $\frac{1}{2}$. Or we can use frequency considerations, like observing 1,000 coin flips and noting that roughly half of them come up heads. But how are these considerations linked to the single and very real event of the next coin flip? What do they really tell us? If we can't answer these questions, then why should we believe in probability theory at all?

One intriguing justification of the theory was developed during the twentieth century and it involves the mathematics of games. The central idea is that probabilities are not objective quantities that are out there in the world, but subjective degrees of belief: if I say the probability of heads on a coin is $\frac{1}{2}$ then that's not because the number $\frac{1}{2}$ is somehow attached to the coin, but because for some reason or other I am 50% confident that heads will come up. These degrees of belief are very personal, but it's possible to measure them using gambling. For example, you could say that my degree of belief in the coin coming up heads is measured by the amount of money (or other currency I value) I am willing to buy or sell a bet on heads coming up at the next flip.

What philosophers have been able to show is that if you assume that a person adheres to some very basic principles of rationality, then they should stick to the rules of probability theory when calculating with their personal degrees of beliefs. If they don't stick to these rules, then they can be made to enter bets in which they will definitely lose money.

As an example, suppose I have decided that my degree of belief of heads coming up on a coin flip is $\frac{1}{2}$ and that the probability of tails is $\frac{1}{4}$. This violates rule number one of probability theory: the probabilities of alternative outcomes should add to 1, but here they only add to $\frac{3}{4}$. Now you can convince me to offer a bet with even odds on heads and 3 to 1 odds on tails – these are the odds that reflect the probabilities I believe so they are acceptable to me. If you now bet £20 on heads and £10 on tails you will take £40 if heads comes up (the £20 you bet and £20 you won) and £40 if tails comes up (the £10 you bet and £30 you won). Since you bet £30 in total you will make a sure profit of £10. So there's guaranteed loss for me.

This, so philosophers have argued, gives some justification to the mathematics of probability. It makes rational sense to stick to it. The obvious objection is that people don't treat their lives as a series of betting games. I could simply have refused to enter any bets and kept my money. But the point of this argument is not to simulate real people, but to give a reason for why we should listen to probability theory at all. It's something to keep in mind next time it rains when you don't have an umbrella. The weather forecast may have turned out wrong yet again, but at least the underlying maths of probability is rationally compelling. Forecasts, by their very nature, cannot be perfect. In fact, nothing can – except, perhaps, for numbers.

16929639 . . . 270130176
The height of perfection

Supposedly, God created the world in six days and on the seventh he had a rest. It's a shame he didn't manage the job a little more quickly, because we could all do with a few more Sundays. But, in any case, it seems that the number of days it took God to create the world was no accident. He chose it for a reason, according to Saint Augustine (354–430) in his famous work *The City of God:*

> *Six is a number perfect in itself, and not because God created all things in six days; rather, the converse is true. God created all things in six days because the number is perfect, and it would have been perfect even if the work of six days did not exist.*

What makes six so perfect? It's an even number, but that in itself isn't enough as there are infinitely many other, equally even numbers. It's divisible by 3, but again, there are infinitely many other numbers that are as well. The only other divisor of 6 (apart from itself) is 1, but that's even less special, since the same goes for every single whole number on the number line.

To see the perfection of 6, you need to add up its divisors, 1, 2 and 3:

$$1 + 2 + 3 = 6.$$

Six is the sum of all its proper divisors (all divisors apart from itself). It combines addition and multiplication to perfection:

$$6 = 1 + 2 + 3 = 1 \times 2 \times 3.$$

If you add up the proper divisors of other numbers you soon find that this property is rare indeed. Many fall short of the target. For example, the proper divisors of 4 are 1 and 2, and

$$1 + 2 = 3,$$

which is less than 4.

Similarly, the proper divisors of 10 are 1, 2 and 5, and

$$1 + 2 + 5 = 8,$$

which is less than 10.

Nicomachus of Gerasa, a prominent member of the Pythagorean school, who were fond of numerology, had a name for such numbers, and it wasn't flattering. He called them *deficient*, producing 'wanting, defaults, privations and insufficiencies'. If they were animals, so Nicomachus wrote, such numbers would be like something with 'a single eye, . . . one armed or one of his hands has less than five fingers, or if he does not have a tongue . . .'

The first number apart from six that isn't deficient is 12. Its proper divisors are 1, 2, 3, 4 and 6, which add up to

$$1 + 2 + 3 + 4 + 6 = 16,$$

which overshoots the target.

However, numbers for which the sum of proper divisors is greater than the original number weren't favourites with Nicomachus either. He called them *abundant*, producing 'excess, superfluity, exaggerations and abuse'. In animal terms they are like creatures with 'Ten mouths, or nine lips, and provided with three lines of teeth; or with a hundred arms, or having too many fingers on one of its hands . . .'

Perfection, he thought, resides in the balance: 'And in the case of those that are found between the too much and the too little, that is in equality, is produced virtue, just measure, propriety, beauty and things of that sort – of which the most exemplary form is that type of number which is called perfect.'

Six isn't alone in its perfection. The next perfect number is 28. Its proper divisors are 1, 2, 4, 7 and 14, and

$$1 + 2 + 4 + 7 + 14 = 28.$$

This, some have suggested, is why it takes the Moon roughly 28 days to move once around the Earth. Perhaps, it's another one of God's careful choices.

Rare and precious

Perfect numbers are rare jewels. Throughout antiquity, and until well into the middle ages, only four perfect numbers were known, 6, 28, 496 and 8,128. In the 13th century the Arab mathematician Ismail ibn Ibrahim ibn Fallus nearly doubled their number by finding another three:

33,550,336

8,589,869,056

137,438,691,328.

European scholars didn't know about his work though, and had to painstakingly rediscover these three perfect jewels. After that, progress remained slow. By 1914 a mere 12 perfect numbers had been discovered, and today, even with the help of fast computers, we still only know of 48 of them. The largest so far, discovered in January 2013, has over 34 million digits, which we don't have space to show here. Suffice to say that it starts with 16929639 and ends with 270130176.

Even perfection

This leads us to the million dollar question: how many perfect numbers are there, and will we ever find more? Nicomachus, although he knew of only four perfect numbers, postulated that an infinite number of them lie concealed within the number line. Mathematicians still believe that this is true, but so far nobody has been able to prove it; we simply can't say for certain how many perfect numbers there are and it's possible that they will one day run out.

The best chance we have to find more perfect numbers hinges on an intriguing pattern that was first discovered by the great Euclid. It links the perfect numbers to that other class of numbers that has fascinated mathematicians since ancient times: the primes.

Stated in modern terms, Euclid had noticed that the first perfect number, 6, can be written in terms of powers of 2:

$$6 = 2 \times 3 = 2^1 \times (2^2 - 1).$$

Something similar works for the other three perfect numbers that were known at the time:

$$28 = 4 \times 7 = 2^2 \times (2^3 - 1)$$
$$496 = 16 \times 31 = 2^4 \times (2^5 - 1)$$
$$8128 = 64 \times 127 = 2^6 \times (2^7 - 1).$$

The pattern is now becoming visible – perhaps the next perfect number is $2^8 \times (2^9 - 1)$? Unfortunately, the answer is no:

$$2^8 \times (2^9 - 1) = 130,816$$

isn't perfect. Only four of its factors 65,408, 32,704, 18,688 and 16,352 add to 133,152, which is more than the number itself, making 130,816 into one of those horrible abundant numbers. But, hang on a second, let's study these products for known perfect numbers a little more closely. In all four examples, the second factor, which has the form $2^k - 1$ where k is either 2, 3, 5 or 7, is a prime number:

$2^2 - 1 = 3$ is prime,

$2^3 - 1 = 7$ is prime,

$2^5 - 1 = 31$ is prime,

$2^7 - 1 = 127$ is prime.

Euclid proved that this isn't an accident: if you can find another prime that can be written as $2^k - 1$ for some whole number k, then the corresponding number

$$2^{k-1} \times (2^k - 1)$$

will be perfect. It's clear that it will be an even perfect number, since you're multiplying by powers of 2. But we can say something even stronger than this, although it was a hard slog to get there. Around 2,000 years after Euclid, Leonhard Euler managed to show that Euclid's recipe is exhaustive: every even perfect number can be written as $2^{k-1} \times (2^k - 1)$, for some positive whole number k, with $2^k - 1$ a prime number. (What an odd perfect number would look like, or if it even exists, we don't know.)

The hunt for perfection

The modern hunt for perfect numbers is based entirely on Euclid's idea and it piggy-backs on the hunt for prime numbers. As we saw in Chapter 2, finding primes is quite a challenge because factoring numbers is incredibly hard work. Things aren't too bad if you are looking for prime numbers that are relatively small. If you are searching for primes within the list of numbers from 1 up to around 1,000,000, you can use a recipe that you might well have met at school, and which was first described by none other than Nicomachus in his *Introduction to Arithmetic*.

You start by listing all those numbers you want to sift for primes, say the numbers from 1 to 1,000,000. Then cross out all the multiples of 2 other than 2 itself. Then you cross out all the multiples of 3, other than 3 itself. The next uncrossed number is 5, so you leave 5 and cross out all the multiples of 5, and so on. If you carry on like this you eventually cross out all the numbers that are multiples of other numbers greater than 1. And since the primes are precisely those numbers that aren't multiples of any number bigger than 1 and smaller than themselves, you end up with only the prime numbers remaining.

Nicomachus attributed this method to Eratosthenes of Cyrene and it now carries the latter's name: the *Sieve of Eratosthenes*. Despite its old age and rather pedestrian nature it is still the most efficient method to find primes up to around $N = 1,000,000$. But it also illustrates the main problem with finding primes: the larger the number N the more steps you need to execute to complete the sieve.

This is why when it comes to finding perfect numbers, we are rather lucky. The numbers that form part of our formula for even perfect numbers, which can be written as $2^k - 1$, are special candidates for prime numbers. There are faster methods for checking if they are indeed primes than there are for other numbers. They are called Mersenne numbers, after the French monk Marin Mersenne who studied them in the middle of the 17th century.

Finding primes on borrowed time

Mersenne numbers are the target of the longest-running grassroots supercomputing project ever launched: the Great Internet Mersenne Prime Search (GIMPS). It started in 1996 and uses computing time donated by volunteers; you download a free program from the GIMPS site and while your computer is idle it combs the number line for Mersenne numbers that are primes. Fourteen of the 48 known Mersenne primes have been found in this way. There's a cash award for anyone whose computer finds a new Mersenne prime, but most people are in it for the thrill of the chase. On 25 January 2013 a computer belonging to Curtis Cooper, a professor at the University of Central Missouri, broke a four-year drought in the hunt: it found the 48th and to date largest Mersenne prime, which is also the largest known prime full stop. It's $2^{57,885,161} - 1$ and has over 17 million digits. And it's exactly the Mersenne prime that gives us the largest known perfect number:

$$2^{57,885,160} \times (2^{57,885,161} - 1).$$

It's hard to tell when the GIMPS team will have cause to break open the champagne again. As the Mersenne numbers involved get larger, it gets harder to check if they are prime and therefore if they give you a perfect number. And there is of course also the more fundamental problem. Nobody has been able to prove Nicomachus' great claim that there are infinitely many perfect numbers or, equivalently, infinitely many Mersenne primes. If there aren't, then we will one day find the very last one and number hunters will have to look for a new toy.

And what of odd perfect numbers? Nicomachus thought that none exist, and although nobody has been able to prove that claim either, it seems that he was right. Mathematicians have combed all the numbers between 1 and $10^{1,500}$ and they haven't found one. They have also shown that if there were one, it would have to satisfy a staggering range of conditions, making its existence seem very unlikely. As the English mathematician James Joseph Sylvester said in 1888, 'A prolonged meditation on the subject has satisfied me that the existence of any one such [odd perfect number] – its escape, so to say, from the complex web of conditions which hem it in on all sides – would be little short of a miracle.'

The hunt for primes and perfect numbers has led us to unimaginable heights – the largest known prime and the largest known perfect number are so big, we cannot possibly conceive of them. But even these giants are dwarfed by the number we will explore next.

Graham's number
Too big to write but not too big for Graham

What's the biggest number you can think of? Infinity you say? Well, which one? We'll open that can of worms in the next chapter. But for now, what's the biggest finite number you can think of? You might start off with some of the mind-bogglingly large numbers thrown up by the physical world. The age of the Universe is 13.77 billion years or 4.343×10^{17} seconds. *Avogadro's number*, the number of hydrogen atoms in 1 gram of hydrogen (called a mole – the standard unit for measuring an amount of a substance in chemistry or physics), is a sizeable $6.02214129 \times 10^{23}$. Bigger than this is the number of atoms in the observable Universe, thought to be between 10^{78} and 10^{82}.

But these are already eclipsed by the 48th Mersenne prime that we saw in last chapter,

$$2^{57,885,161} - 1.$$

This is the biggest prime number we know of and has an impressive 17,425,170 digits. It is also bigger than the famous *googol*, 10^{100} (a 1 followed by 100 zeros), defined in 1929 by American mathematician Edward Kasner and named by his nine-year-old nephew, Milton Sirotta. Milton went even further and came up with the *googolplex*, now defined as $10^{googol} = 10^{10^{100}}$ but initially defined by Milton as a 1, followed by writing zeros until you get tired.

A googolplex is significantly larger than the 48th Mersenne prime. You, or rather a computer, can write out the 48th Mersenne prime in its entirety, all 17,425,170 digits of it. But, despite the fact that I can tell you what any digit in the googolplex is, no person, no computer, no civilization will ever be able to write it out in full. This is because there is not enough room in the Universe to write down all googol digits of a googolplex. As Kasner, and his colleague James Newman, said of the googolplex (in their 1940 book *Mathematics and the Imagination*, which introduced the world to these numbers): 'You will get some idea of the size of this very large but finite number from the fact that there would not be enough room to write it, if you went to the farthest star, touring all the nebulae and putting down zeros every inch of the way.'

In 1970s, however, there appeared a huge number that dwarfed all that had come before it. But in order to understand where this gargantuan of maths appeared from, we need to first go to a party, and then get very good at our 3 times table.

Friends and strangers

Deciding on a guest list is always tricky. You're pretty sure your piano teacher, Grigori, will not know anyone else. But your school friends from calculus class – Isaac, Gottfried, Robert and Leonard – are thick as thieves, all knowing each other. And your best friend Emmy, who was great at spotting similarities between people, always knows a like-minded party-goer. Exactly how many people are friends and how many people are strangers in a group can make or break a party.

To keep track of it all, suppose you drew map of the relationships of all your friends, linking two people with a black edge if they were friends and with a grey edge if they were strangers. Suppose your friend Ada doesn't know your friends Benoît and Claude but they know each other:

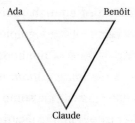

Your other friend David knows Ada and Claude, but doesn't know Benoit:

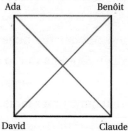

And if you add in your friends Emmy and Florence you have:

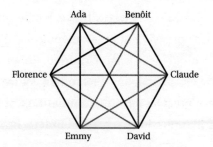

You might recognize these pictures from Chapter 6: they are *networks* where the nodes are your friends and the links are their relationships. Two things make these networks special. The first is that they are *fully connected* – every person is connected by a link to every other person in the network. Secondly, the links in these relationship networks are coloured – grey for strangers and black for friends.

There are obviously lots of different ways a relationship network could be coloured (in fact 2^n ways for n edges). But if you look closely at a larger network, you might find it hides simpler examples, for example a smaller group of people whose links are all the same colour. This represents a group that are all friends or all strangers. In our example above, Benoit, David and Emmy are all strangers, forming a grey triangle. And once we added Florence into the mix, we also have a black triangle as Ada, Emmy and Florence are all friends.

No such simple examples appear in the relationship network between your first four friends: Ada, Benoit, Claude and David. It was only when you added Emmy into the mix that a group of three people who were either all friends or all strangers appeared. In fact it seems that as you include more and more people in the network, smaller one-coloured groups start popping up.

Getting your relationships in order

A brilliant young mathematician, Frank Ramsey, noticed the same thing in 1928, though he wasn't planning a party – he was working on a paper about logic. In particular, Ramsey was interested in when it was possible to guarantee a certain amount of order to be present in a system.

You can think of these simpler one-coloured groups of all friends or all strangers as pockets of order arising in the larger network. The more ordered the network the larger these one-coloured groups would be, until we get to the most ordered state possible where every link in the whole network is the same colour. An ordered system is one in which everything obeys some sort of rule, such as 'every link is grey'. Therefore ordered systems are much simpler than disordered systems, in terms of the information needed to describe them. The relationship network of Benoît, David and Emmy is a triangle with all three sides coloured grey, much simpler to describe than that for Ada, Benoît and Claude, for which the colour of each side and the people it links must be specified.

Ramsey proved what we have begun to suspect: no matter how much order you want, with a big enough network you can guarantee it will appear. Following our example above, you might suspect that you are guaranteed to find a group of three people who are all friends (black triangle) or all strangers (grey triangle) if your relationship network contains six or more people. And you'd be right. The proof is surprisingly straightforward and in essence no matter what amount of order you are looking for (that is, no matter what size of one-coloured group you want consisting of all friends or all strangers), if your network is large enough you can guarantee that it will contain that much order.

Ramsey's proof of this was just one step on the way (something mathematicians call a *lemma*) to the main result he was working towards in his paper on logic. This one stepping stone in his proof spawned *Ramsey theory*, a whole new area of research in combinatorics (loosely, the maths of counting stuff). This is an example of Ramsey's brilliance, creating a whole new area of maths as a by-product of something else he was working on. But Ramsey was not just a brilliant mathematician. As well as contributing to the areas of mathematics and logic, he also published influential papers in economics and philosophy, all before he died of jaundice at the terribly young age of 26.

Ramsey's numbers

It's fairly straightforward to prove that six is the smallest number of people you need to guarantee that there will be three people who are all friends, or three who are all strangers. In the language of Ramsey theory this is called a *Ramsey number*, $R(n, m)$ – the smallest number of people you need to guarantee that you have a group of n people who are all friends (that is, n nodes all connected by black edges) or a group of m people who are all strangers (that is, m nodes all connected by grey edges).

In fact there's a whole family of Ramsey numbers that we can work out without much effort: those like $R(2, m)$, the number of people you need to guarantee there are either 2 friends or m strangers. And the answer is always m people. If any two people are friends you are done. Otherwise they are all strangers, and you are done again.

The other piece of information you get for free is that if you know $R(n, m)$, then you immediately know $R(m, n)$: this is because $R(n, m) = R(m, n)$. The colour you paint your lines doesn't matter, if you can guarantee you have n friends and m strangers, then you can just swap the colours of the lines over and guarantee that you have m strangers and n friends.

But things quickly get tricky when you start looking for groups of more than three people who are all friends or all strangers. In fact, if n and m are both bigger than 2 (so we can't use our simple trick above) then we only know the following Ramsey numbers:

$R(3, 3) = 6,$

$R(3, 4) = R(4, 3) = 9,$

$R(3, 5) = R(5, 3) = 14,$

$R(3, 6) = R(6, 3) = 18,$

$R(3, 7) = R(7, 3) = 23,$

$R(3, 8) = R(8, 3) = 28,$

$R(3, 9) = R(9, 3) = 36,$

$R(4, 4) = 18,$

$R(4, 5) = R(5, 4) = 25.$

So mathematicians have managed to go one better than our initial example, and have proved that the number of people needed to guarantee that there are four people who are all friends or four who are all strangers is 18 (that is, $R(4, 4) = 18$). But that's as far as they've got. Even for a number as small as 5, we can't say how many people we need in our graph to guarantee five people who are all friends or all strangers.

How parties quickly get out of hand

Just like a party can quickly get out of hand if too many people turn up, part of the problem with Ramsey numbers is how quickly the number of edges in your networks grows as you include more people. With a group of three people, say Ada, Benoît and Claude, we only had to think about three different relationships: whether Ada knew Benoît, whether Benoît knew Claude and whether Claude knew Ada. If we add in another person, David, to take us up to four people, we have six different relationships to consider.

As we noted above, these relationship networks are fully connected, that is, every person is connected to every other person. The number of edges in these fully connected networks (also known as *complete graphs*) quickly escalates as you increase the number of people. And the number of possible colourings grows even faster.

One way to calculate Ramsey numbers would be to use brute force to check each of the possible colourings of graphs with a particular number of nodes. But this isn't a feasible approach for even the smallest numbers of friends and strangers, such as the Ramsey number $R(3, 3)$ which we know is equal to six. To prove this result using brute force we would have had to have checked each of the $2^{15} = 32,768$ coloured graphs of 6 people to make sure they all had a black or grey triangle representing 3 friends or 3 strangers. Even using symmetry to reduce the workload (only checking once for things that are just symmetric images of others, for example, with the colours of edges swapped or for networks that are mirror images of each other) it is still an unreasonable approach.

Number of people/nodes	Number of edges/relationships	Number of possible colourings of this graph
2	1	2
3	3	$2^3 = 8$
4	6	$2^6 = 64$
5	10	$2^{10} = 1{,}024$
6	15	$2^{15} = 32{,}768$
.	.	.
.	.	.
.	.	.
n	$\dfrac{n(n-1)}{2}$	$2^{\frac{n(n-1)}{2}}$

Instead mathematicians often use a counter example to establish a lower bound. Although we don't know the exact value of $R(5, 5)$, we do know that it has to be greater than 42. This is because Geoffrey Exoo, a professor of computer science at Indiana State University, found a counterexample: a particular graph of the relationships between 42 people that did not have a group of 5 people who were all friends or all strangers.

The next step is to establish how many friends is enough to guarantee your groups of friends or strangers. Exoo and fellow mathematicians Brendan McKay from Australian National University and Stanislaw Radziszowski from the Rochester Institute of Technology were able to show that 49 people was enough to guarantee that you have a group of 5 people who were either all friends or all strangers. So we know for certain that the Ramsey number, $R(5, 5)$ is somewhere between 43 and 49.

There is strong evidence that $R(5, 5)$ is indeed equal to 43. But a definitive proof of this has so far eluded mathematicians and computer scientists alike. And the days of checking by brute force are long gone: checking every possible coloured graph on 43 people would mean considering $2^{43 \times \frac{42}{2}}$ possible graphs. This is more graphs than there are atoms in the observable universe. Suddenly hunting for a needle in a haystack seems a far more achievable pastime.

Graham's big number

We might complain about only being able to narrow down $R(5, 5)$ to the range of 43 to 49. But that, in the wider landscape of Ramsey theory, is nothing. In 1971 the American mathematician Ronald Graham unveiled an upper bound on a particular problem in Ramsey theory that puts a range of merely seven numbers into the shade.

Graham was used to juggling big numbers. In fact, he'd been in the circus while he was studying and maintained his regard as one of the best jugglers in America by having a specially constructed net hung from the roof of his office in Bell Laboratories. The net had a hole in the centre, which he drew tightly around his waist so that when he dropped one of the six or seven balls he was juggling it would obligingly roll back to him.

Graham and his colleagues were working on Ramsey numbers for a special sort of relationship network. Imagine a network for eight people, with each edge as usual coloured either grey or black. But instead of the eight points lying flat on the paper, they are the eight corners of a cube. Then each face of the cube represents four people, all linked by grey or black edges running across the face of the cube. You could also slice the cube diagonally from one edge to another, giving four people linked by six grey or black edges. Graham wanted to know when the network was coloured in such a way that all the flat planes of the cube containing four corners (the faces or diagonal slices) contained both grey and black edges: that none of the planes of the cube had all their edges of just one colour.

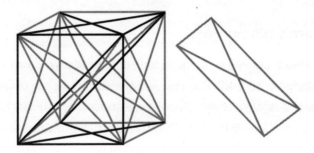

A cube with all its corners joined by black or grey lines. The diagonal plane slicing through the centre of the cube, shown here, is an example of four corners, lying on a plane, that are all joined by the same colour of lines. By changing the bottom edge of this slice, from grey to black, you would create a counterexample, as there would then be no plane through the cube, containing four corners, that had all lines of the same colour.

Now remembering those higher dimensions from Chapter 4, imagine that you instead start with a four-dimensional cube: this has 16 corners (rather than the 8 of a three-dimensional cube) and these are all joined to each other. Or that you started with a five-dimensional hypercube, with its $2^5 = 32$ corners all joined, or indeed an n-dimensional hypercube with its 2^n corners all joined. Graham wanted to know what dimension of hypercube you need to guarantee that, for every colouring, there will always be four corners, all on a plane, joined by lines of the same colour.

Perhaps unsurprisingly, nobody has been able to say exactly what dimension is needed. But Graham did manage to give an explicit upper bound on the problem, in a similar way to the upper bound on some of the Ramsey numbers earlier, such as the upper bound of 49 for $R(5, 5)$. The difference, however, is that Graham's upper bound is absolutely huge. It is so huge that it is impossible to write in normal notation. In fact, just as for the googolplex (which is significantly smaller than Graham's number), even if you could write each digit in tiny Planck-length sized writing, these digits would take up more space than the observable Universe. It is absolutely bloody enormous.

Writing the unwritable

What is amazing about Graham's number is that even though it is so huge we could never hope to ever write it, we are able to state the number exactly. That's because we can use an ingenious shorthand that is not dissimilar to the tricks of scientific notation we examined in Chapter 10 that made it easier to express extremely large or small numbers.

Graham's number can be exactly expressed using notation developed by the mathematician and computer scientist Donald

Knuth. Knuth's *up-arrow notation* continues the compounding nature of the better-known arithmetic operations. As we saw in Chapter 1, multiplication is simply repeated addition:

$$a \times b = \underbrace{a + a + \dots + a}_{b \text{ times}}.$$

So, for example, $3 \times 3 = 3 + 3 + 3 = 9$.

Similarly we saw in Chapter e that exponentiation, or raising a number to the power of another number, is just repeated multiplication:

$$a^b = \underbrace{a \times a \times \dots \times a}_{b \text{ times}}.$$

So this time our example might be $3^3 = 3 \times 3 \times 3 = 27$.

Knuth developed an ingenious system that just allowed this process to carry on, defining infinitely many more levels of arithmetic operations. The first step was another way of expressing the familiar exponentiation. Knuth denoted this repeated multiplication with a single up-arrow, ↑:

$$a \uparrow b = \underbrace{a \times a \times \dots \times a}_{b \text{ times}} = a^b.$$

So repeating our example we have $3 \uparrow 3 = 3^3 = 27$.

Then the next operation was repeated exponentiation, sometimes called a *power tower*, and notated with a double up-arrow ↑↑:

$$a \uparrow\uparrow b = \underbrace{a \uparrow (a \uparrow (\dots \uparrow a))}_{b \text{ copies of } a} = a^{a^{a^{\cdot^{\cdot^{\cdot^a}}}}}.$$

For example, a power tower that is 3 levels high is:

$$3\uparrow\uparrow3 = 3\uparrow(3\uparrow3)) = 3^{3^3} = 3^{27} = 7{,}625{,}597{,}484{,}987.$$

This definition can be repeated indefinitely, allowing us to easily express progressively larger and larger numbers:

$$a \uparrow\uparrow\uparrow b = \underbrace{a \uparrow\uparrow (a \uparrow\uparrow (... \uparrow\uparrow a))}_{b \text{ copies of } a}.$$

So continuing our example:

$$3\uparrow\uparrow\uparrow3 = 3\uparrow\uparrow(3\uparrow\uparrow3) = 3\uparrow\uparrow(3\uparrow(3\uparrow3)) = 3\uparrow\uparrow7{,}625{,}597{,}484{,}987,$$

which would be a power tower of 3s that was 7,625,597,484,987 levels high. We'll leave you to calculate its exact value as an exercise for some insomnia-blighted night.

This has already got out of hand from the point of view of numbers that we can actually comprehend. But just as mathematicians are quite happy thinking in arbitrarily high dimensions if the rules are clearly defined, they are quite happy to continue Knuth's nice, well-defined process of iterating the arithmetic operations.

Too big to write, but it ends in 7

When Graham was explaining his result to the mathematical enthusiast and general legend Martin Gardner (whom we met in Chapter 7), Graham fudged a little, giving a (relatively slightly) larger number (and so also an upper bound on the problem) that was a tiny bit easier to state. This is the number that has come to be known as *Graham's number* and you'll find it in the box opposite. As Knuth said the dots 'suppress a lot of the details'.

GRAHAM'S NUMBER

To specify what Graham's number is exactly, we need to go through a process of 64 steps. The first step is to define a number, which we'll call g_1, using the up-arrow notation:

$$g_1 = 3 \uparrow\uparrow\uparrow\uparrow 3$$

Then the next step is to define a number g_2,

$$g_2 = 3 \uparrow\uparrow\uparrow\uparrow \ldots \uparrow\uparrow\uparrow\uparrow 3,$$

where the number of up-arrows here is g_1.

Then we carry on in this way. The next number g_3,

$$g_3 = 3 \uparrow\uparrow\uparrow\uparrow\uparrow\uparrow\uparrow\uparrow \ldots \uparrow\uparrow\uparrow\uparrow\uparrow\uparrow\uparrow\uparrow 3,$$

has g_2 up-arrows, and g_4,

$$g_4 = 3 \uparrow\uparrow\uparrow\uparrow\uparrow\uparrow\uparrow\uparrow\uparrow\uparrow\uparrow\uparrow \ldots \uparrow\uparrow\uparrow\uparrow\uparrow\uparrow\uparrow\uparrow\uparrow\uparrow\uparrow\uparrow 3,$$

has g_3 up-arrows, and so on. This continues for 64 steps, to give

$$g_{64} = 3 \uparrow \ldots \uparrow 3$$

which has g_{63} up-arrows. Now that, my friends, is a big number.

Graham published his upper bound in a paper in 1971, and it was then the largest number ever explicitly used in a mathematical proof. It was only marginally smaller than the number that Gardner published in his *Scientific American* column, which came to be known as Graham's number. At the time of his paper, the lower bound for the problem involving hypercubes was just 6, so Graham had proved that the answer was somewhere between 6 and his giant record-breaking number. Graham, in a triumph of understatement, suggested that 'Clearly, there is some room for improvement here.'

Graham must be pleased with the progress made since then. In 2013 the upper bound has been significantly reduced to $2\uparrow\uparrow\uparrow\uparrow6$. And various counterexamples have been found to improve the lower bound to 11. However, it is still not clear which of the gaps, either the gap of 43 to 49 for the ordinary Ramsey number $R(5, 5)$, or the gap of 11 to $2\uparrow\uparrow\uparrow\uparrow6$, for the hypercube problem, will be closed first.

What is even more mind-boggling about Graham's number is that, even though it is so ginormous that the Universe is not big enough to write the number out, we do know that the last digit is 7. In fact, we know about 500 of the last digits. (This is because the right-most digits of any power tower end up remaining the same as you raise the tower to further and further powers.) It's reassuring that although Graham's number is far too big to make any sense, we have its tail under control – something that cannot really be said for the next 'number' we are going to look at.

∞ Are we there yet?

nfinity is a strange thing. Everyone has an instinctive sense of what it is, but things get tricky when you try to pin it down exactly. Luckily, there is a lot we can learn from numbers. Think of the natural numbers

1, 2, 3, 4, 5, 6, . . .

There isn't a largest natural number because whatever number you come up with I can add 1 and get an even larger number. This means you can never get to the end of the natural numbers. They are an example of what Aristotle called a *potential infinity*: something, a list or an expanse, that has no bounds, that you can never get to the end of. It's an infinity alright, but one you will never actually see directly because you can't get there. It's an infinity that doesn't come to bite you.

But still, strange things can happen even with this toothless kind of infinity. Let's take all the even numbers. Every other number is even, so there should be half as many even numbers as natural ones. And this is definitely true if you look at all the even numbers in the sequence from 1 up to, say, 100. It contains 100 numbers in total and 50 of them are even. If you go up to 101, then slightly less than half are even, $\frac{50}{101} = 0.495$. But that's only because you've chosen to end on an odd number. No matter what number you go up to, 1,000, 10,001, 100,025, 10^{10}, the proportion of even numbers will always be $\frac{1}{2}$ or very close.

But now, let's look at *all* the even numbers. There are infinitely many, but we can still list them in order: 2 is the first even number, 4 is the second even number, 6 is the third even number, and so on, giving the following list:

Place on list	Even number
1	2
2	4
3	6
⋮	⋮
20	40
⋮	⋮
n	$2n$
⋮	⋮

That list goes on forever, with every even number associated to the natural number that gives its place on the list. But hang on a second. Every even number is associated with exactly one natural number. Conversely, every natural number is associated with exactly one even number. Now in everyday life if you have two collections of things (say a group of people and some chairs) and you can associate to every thing in one collection exactly one thing in the other (so that's one chair for each person and one person for each chair), then the two collections have the same size: there are as many chairs as there are people.

Applying the same kind of reasoning to our numbers, we see that there are just as many even numbers as there are natural numbers. That's very odd, since intuitively it's crystal clear that even numbers only make up half of all natural numbers. What's worse, you can list all the fractions (the rational numbers) in a similar way (see the argument in the box on page 266). This would imply that there are just as many even numbers as there are natural numbers as there are rational numbers. Ridiculous!

The obvious conclusion is that the business of comparing sizes simply doesn't work for infinite collections of things. Galileo Galilei shared that sentiment. In his 1638 book *Two New Sciences* he wrote,

> *The attributes 'equal', 'greater', and 'less', are not applicable to infinite, but only to finite, quantities.*

Infinity times 2

That might have been that – if it hadn't have been for a mathematician by the name of Georg Cantor. Born in Russia in 1845 Cantor moved to Germany with his parents aged 11 and never quite felt at home in his new country. He did, however, prosper mathematically and eventually produced an analysis of infinity that made a major splash – what mathematician can boast the title 'corrupter of youth' on account of his work?

In the 1870s Cantor was playing similar games to the one we just played, only instead of counting even numbers or rational numbers, he was trying to count all of them. As we saw in Chapter $\sqrt{2}$, there are also irrational numbers that can't be written as fractions, for example $\sqrt{2}$, e, π, ϕ and many more. If you add them to the game, then you are looking at *all* the numbers that live on the number line – the *real numbers*. Each of them is represented by its decimal expansion, which can be finite or infinite. (Strictly speaking we have to account for ambiguities, such as $0.999999... = 1$, but that's not too hard to do.)

COUNTING THE FRACTIONS

In between two natural numbers, say 1 and 2, there is always a fraction, for example $1\frac{1}{2} = \frac{3}{2}$. Actually there are many more fractions in between any two natural numbers. The natural numbers can themselves be written as fractions, for example $2 = \frac{2}{1}$. So clearly there are many more rational numbers (those that can be written as fractions) than there are natural numbers.

If you try to list the (positive) rational numbers in order, as we did with the even numbers, you soon run into trouble because you won't know where to start. If you start the list with $\frac{1}{2}$, then you have missed all the fractions smaller than $\frac{1}{2}$, such as $\frac{1}{4}$, $\frac{3}{7}$ or $\frac{1}{100}$. If instead you start with $\frac{1}{100}$ then you've missed out all fractions smaller than that, such as $\frac{1}{101}$, $\frac{1}{1,000}$, or $\frac{5}{22,222}$. There isn't a smallest positive rational number to start your list with.

But who said that you need to list the fractions in order of size? Let's instead use another recipe. First list all the fractions for which the denominator and the numerator add up to 2: there's only one, namely $\frac{1}{1} = 1$. Now list all the fractions whose denominator and numerator add up to 3: that's $\frac{1}{2}$ and $\frac{2}{1}$ and we can list those two by size. Next are those where the sum is 4: that's $\frac{1}{3}$, $\frac{2}{2}$ and $\frac{3}{1}$. We ignore $\frac{2}{2}$, because that's equal to 1, which we have already counted. Continuing like this gives us an unfailing recipe to list *all* the positive rational numbers:

Place on list	Rational number	Numerator and denominator add to
1	$\frac{1}{1} = 1$	2
2	$\frac{1}{2}$	3
3	$\frac{2}{1}$	3
4	$\frac{1}{3}$	4
5	$\frac{3}{1}$	4
6	$\frac{1}{4}$	5
7	$\frac{2}{3}$	5
8	$\frac{3}{2}$	5
9	$\frac{4}{1} = 4$	5
$\vdots$	$\vdots$	$\vdots$

As with the rational numbers, we can't list the positive real numbers by size. Coming up with a recipe for listing them one by one is going to be pretty tricky. So let's simply assume, for the sake of the argument, that we have found such a recipe and that we have constructed a list of positive real numbers. Just for illustration, let's assume it starts like this:

Place on list	Real number
1	0.12
2	2.543
3	4
4	3.123456...
5	0.3333...
6	100.67
⋮	⋮

Now we will make a new number as follows. It starts with a zero followed by the decimal point. For the first digit after the decimal point, add 1 to the first digit after the decimal point in the first number in the list (if you get 10, replace this with the digit 0). In our example, the first number in the list is 0.12, so adding 1 to 1 we get 2 and our new number so far is

0.2.

For the second digit after the decimal point, repeat this process with the second digit after the decimal point of our second number. In our example, the second number in the list is 2.543 and the second digit is 4, so we add 1 to get 5. Our new number is now

0.25.

For the third digit after the decimal point, repeat the process with the third digit after the decimal point of our third number. That seems impossible at first, since 4 doesn't have any numbers after the decimal point, but it's OK if we notice that $4 = 4.00000\ldots$ So the third digit after the decimal point is 0, and adding 1 we get 1. The number now is

0.251.

After the next step it will be 0.2515, after the next 0.25154, then 0.251541, and so on. You can keep going like this forever (well, at least in theory) giving you a perfectly respectable real number in the end.

This real number should appear somewhere on our list if we have indeed listed all positive real numbers. Let's try to find it. It can't be the first number on our list, because the two differ in the digit after the decimal point (since we added 1 to it to get our new number). It can't be the second number on our list, because the two differ in the second digit after the decimal point. For the same reason it can't be the third, the fourth, the fifth, the sixth – it can't be on our list at all! This means that our list was incomplete, missing out at least one number.

The same argument shows that *any* list of positive real numbers must be incomplete. That seems fair enough: intuitively it's clear that there should be more real numbers than natural ones, so we shouldn't be able to match them up exactly. But think again and you'll see that we have just found a second type of infinity. The first is represented by the natural numbers and the second by *all* the positive real numbers. And the latter, as we've just seen, is most definitely bigger than the former. So it *does* make sense to compare the sizes of infinite things!

Towering infinity

It was Cantor who, in 1891, showed that the positive real numbers can't be counted in the same way as the even numbers or the rational ones. And he decided to take on board the fact that infinite collections can have different 'sizes' or *cardinalities*, as he called them. Any collection that has the same cardinality as the natural numbers is called *countably infinite*, for obvious reasons. As we saw above, both the even numbers and the rational numbers are countably infinite.

Cantor himself was quite shocked by what he found using his matching-up arguments. It was he who discovered, in 1877, that you could exactly match the points inside a square to the points in one of its sides. He was stunned by this counter-intuitive idea: 'I see it, but I don't believe it,' he wrote. And it's clear why – the square feels so much bigger than one of its sides, yet in terms of cardinality, as defined by whether you can match things up exactly, the two are the same.

But what is even more surprising is that Cantor managed to construct a whole tower of infinities, one bigger than the other. The smallest one is the countable infinity, as represented by the natural numbers, which he gave the name $\aleph_0$ (pronounced 'aleph-nought'; aleph is the first letter in the Hebrew alphabet). The second type of infinity he called $\aleph_1$, the third, $\aleph_2$, and so on. $\aleph_1$ is bigger than $\aleph_0$ in the matching-up sense we described above: when you match every object in a collection of things that has cardinality $\aleph_0$ to one object in a collection of things that has cardinality $\aleph_1$, then there will always be some things left over in the latter.

There are infinitely many of these alephs, that is, the tower of infinities is infinitely high. We can even get a sense of why this must be the case. For illustration, let's look at a finite collection of things, namely the set {1, 2, 3}, made up of three numbers. Any such collection of things can be broken up into smaller collections, called *subsets*, in various ways. In our example, the set {1, 2} is a subset of {1, 2, 3}, and so is the set {2, 3}. In fact we can list all the subsets of the set {1, 2, 3}. They are:

{1}   (the set that has just the number 1 in it),

{2},

{3},

{1, 2},

{2, 3},

{1, 3}.

With mathematical pedantry, we also include the empty set { }, which has nothing in it, and the set {1, 2, 3} itself. So altogether the set {1, 2, 3} has eight subsets. More generally you can show that a finite set of size n has 2^n subsets (in our case $n = 3$).

Now the subsets of {1,2,3} are themselves just 'things'. If you put them together they form their own collection – their own set, called the *power set* of {1,2,3}. As we have just seen, the power set of {1,2,3} contains eight things, while {1,2,3} contains just three. More generally, the power set of a set always contains more things than the original set.

Cantor showed that the same works for infinite sets. The power set of an infinite set always has a bigger cardinality than the original set. So starting with one infinite set, you can make a whole sequence of progressively bigger ones, simply by taking power sets.

One nice surprise is that the power set of the natural numbers has the same cardinality as the real numbers! But this raises a question: is there an infinity in between the two? A collection of things whose cardinality is greater than that of the natural numbers but smaller than that of the real numbers? The answer is that no one knows. Not just because no one has found out yet, but because in a strict mathematical sense it's impossible to know. As we saw in Chapter 1, if mathematics is considered as an axiomatic system, based on a set of basic rules and logic, then there will always be statements which you can't prove true nor false. The question above, known as the *continuum hypothesis*, is an example of one that is undecidable in this way, at least within the accepted axioms of mathematics.

His ideas on infinity earned Cantor a lot of flack from some of his contemporaries. It was Leopold Kronecker, whose lectures Cantor had heard when at university in Berlin, who called him a corrupter of youth, a renegade and also a scientific charlatan. Poincaré thought Cantor's ideas were a grave mathematical malady, a perverse illness that would one day be cured. But the 'malady' never did get cured. Today no one views it this way: Cantor's ideas are an accepted part of mathematics and no one bats an eyelid. The great mathematician Hilbert hailed Cantor's work as a mathematical paradise. And Bertrand Russell thought Cantor was one of the greatest intellects of the 19th century.

Infinity on a page

The endless line of natural numbers is the first and most natural encounter most of us have with potential infinity. But you can also go the other way, creating an infinity that doesn't stretch off into the horizon. Take a square, and divide it up into nine smaller squares of equal size. Now erase the inside of the middle square, leaving behind its sides. You are left with eight squares and you do the same for each of those: divide each of them up into nine smaller squares and erase the inside of the middle one. Do the same again to each of the $8 \times 8 = 64$ little squares you are left with. Keep going, deleting middle squares ad infinitum (although you can't really do this in practice, you can imagine it). What will you be left with in the end?

Clearly the thing you are left with, let's call it S, is not nothing: all the lines that formed sides of squares are still there as part of S, because they were never deleted. In fact, S opens a door into the beautiful world of the infinite. If you zoom in on any of the smaller squares that came up during the construction, but wasn't a central one and so hasn't been deleted, then what you see is *exactly the same* as the whole picture. That's because what happened to the smaller square is exactly what happened to the original one: it got subdivided into nine squares, middle one deleted, and so on, ad infinitum. S is infinitely intricate; as you zoom in on smaller and smaller pieces you never come to an end, as the same picture emerges again and again. It's an infinity you can hold in the palm of your hand.

This feature of S is called *self-similarity*. You might have already guessed where we are heading here: S is a *fractal*. It is called that

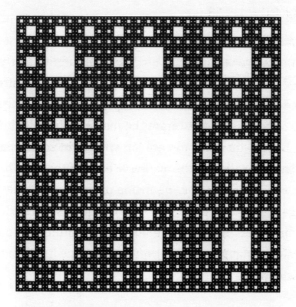

The Sierpinski carpet – a fractal first described by
Wacław Sierpinski in 1916.

because it defies our ordinary concept of dimension. It's not a two-dimensional shape because it covers no area: the original square has been so riddled with holes, no area is left at all in the end. But it's not a one-dimensional object either: you simply can't break it up into a collection of isolated lines or smooth curves. You can convince yourself, for example, that arbitrarily close to the side of a square (which didn't get deleted and so is part of S) are infinitely many sides of smaller squares, which are also part of S.

Boxing it in

How, then, can we characterize such a strange beast? The best way is to erase S from your memory for the moment and think of much simpler shapes, for example a piece of line and a square, sitting side by side on a piece of paper (the black line and the black square, below). To make things easier, imagine that both the piece of line and the sides of the square are 1 cm long. You can fit your line segment neatly into 2 squares whose sides are $\frac{1}{2}$ cm long, sitting side by side (the grey squares covering the black line below). But as you can see, to cover your square by little squares of side length $\frac{1}{2}$ cm, you will need 4 little grey squares.

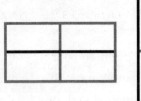 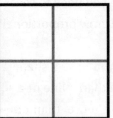

Similarly, you can fit the line segment neatly into 3 squares whose sides are $\frac{1}{3}$ cm long, sitting side by side, while you will need 9 such little squares to cover the original square.

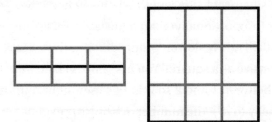

If the little squares have side length $\frac{1}{4}$ cm, you will need only 4 for the line segment, but 16 for the square.

If you have spotted a pattern here then you are correct. To cover the line segment in little squares of side length $\frac{1}{n}$, where n is some positive whole number, you will need n such little squares. But to cover the square in little squares of side length $\frac{1}{n}$, you will need n^2 such little squares. This is in tune with the dimension of the shapes: a line segment has dimension 1, and the exponent of n in the corresponding expression is 1. A square has dimension 2, and the exponent of n in the corresponding expression is 2. Something similar will be true even if the line and sides of the square are longer or shorter: in this case the minimum number of little squares of side length $\frac{1}{n}$ you need to cover the shapes won't be n or n^2 exactly, but they will grow proportionally to n or n^2 as n gets larger.

But what if you have a shape X drawn on your page that is more complicated than a line or a square? What if, like in our example S earlier, X is 'more' than one-dimensional, but 'less' than two-dimensional? You can again cover it in little squares – this gives you a coarse-grained, pixelated approximation of X, which gets better the smaller the squares are. If X is 'more' than one-dimensional but 'less' than two-dimensional, then the minimum number of squares of side length $\frac{1}{n}$ you will need to cover it will grow faster (proportionally) than n (because X isn't just a simple line), but it will grow slower, proportionally, than n^2 (because X isn't as extensive as a square). You may find, in fact, that there is a number d between 1 and 2 so that this minimum number grows proportionally to n^d – the number d, being the exponent of n, can be regarded as the dimension of X. This, loosely speaking, is how you calculate the *box-counting dimension* of a shape.

So what's the box-counting dimension of our shape S, made from perforating squares? It turns out to be approximately 1.8928, which is between 1 and 2, as expected. This is the origin of the word 'fractal': it's a shape whose dimension is fractional, that is, it's not a whole number.

It was the Polish mathematician Wacław Sierpiński who studied *S* in 1916 – it's called the Sierpiński carpet in his honour. You can play the same game by starting with a line segment, rather than a square: erase the middle third of the line, then the middle thirds of the remaining two line segments, then the middle thirds of the surviving four and so on. The resulting *Cantor set*, named after our friend Georg Cantor who studied it, is neither zero nor one-dimensional but has box-counting dimension around 0.63. Moving up to three dimensions you can start with a cube, divide it into 27 smaller cubes and remove some of them, ad infinitum. The resulting *Menger sponge* (named after Karl Menger who constructed it in 1926) has box-counting dimension around 2.726833.

Fractals everywhere

When the first fractals were studied by Cantor, Sierpiński and others, they were considered strange mathematical oddities – and they weren't called fractals. But a few decades later, in the 1960s, an IBM researcher by the name of Benoît B. Mandelbrot made a curious discovery. He was looking at errors in telephone transmissions which, although random, tended to come in bursts. When he looked at such an error burst more closely, he found that it didn't consist of a continuous error that lasted for, say, five minutes, but of smaller error bursts intermingled with error-free periods. Zooming in on smaller error bursts, a similar error and non-error mixture would become apparent – just as you find a mixture of deleted and non-deleted squares when you zoom in on the Sierpiński carpet.

That kind of repeating structure wasn't new to Mandelbrot. He had seen it in the fluctuations of commodity prices, which seemed to vary in similar ways whether you looked at them over the period of a year or a week. And indeed, once you have honed your eye, you find self-similarity everywhere. Our lungs and blood vessels reveal intricate detail at small scales, the purpose being to cover as much space as possible within the confines of our bodies. Earthquakes and storms look the same whether they are big or small, as the laws of physics that govern them don't care about scale. And coastlines look just as crinkly when viewed from close up as they do from far away.

Mandelbrot realized that in many situations it's not size that matters, but patterns that repeat on all scales. It's a complexity that classical geometry is not equipped to deal with. 'Nature exhibits not simply a higher degree but an altogether different level of complexity,' he said. 'The existence of these patterns challenges us to study these forms that Euclid leaves aside as being "formless", to investigate the morphology of the "amorphous".'

Mandelbrot found that the tools developed by those early fractal pioneers exactly fitted the bill. Searching the annals of scientific literature, he found many more examples of fractal-like patterns in nature, and eventually published his famous book *The Fractal Geometry of Nature* in 1982. He made the important link between fractal geometry and dynamical processes: the idea that even simple rules, like those we used to construct the Sierpiński carpet, can give rise to astonishing complexity and, in some dynamic processes, unpredictability. *The Fractal Geometry of Nature* is a seminal work not only in geometry, but also in what has become known as chaos theory.

Mandelbrot was well aware that his insights were revolutionary – too much so for some of his contemporaries, who would have preferred a little more modesty on his part. Cantankerous and eccentric, he always remained a maverick who constantly switched fields. His efforts did get rewarded, however. By the time he died in 2010 he had been showered with prizes and awards, fractal geometry had become an accepted field of mathematics, and his name will live on in one of the most famous fractals of them all, the *Mandelbrot set*. And rumour has it that as well as naming a fractal he also had a fractal name: some say the B. in Benoît B. Mandelbrot stands for Benoît B. Mandelbrot!

Infinities in nature

Fractals embody infinity in their infinite detail, but really they are only imaginary. You could never actually print out a Sierpiński carpet on a bit of paper, because no printer is capable of such detail. And that may not be the only handicap – as we saw in Chapter 10, some physicists believe that you can't zoom in on space to an arbitrary level of precision because there's a smallest fundamental distance. So, in that sense, true fractals with infinite intricacy may exist only in our mathematical imagination.

But are there *any* infinities in nature? The one that first springs to mind is the potential infinity of the Universe. Whether or not it goes on forever is something we simply don't know. It's hard to imagine a finite Universe, because if the Universe is all there is, and it's finite, then what surrounds it? But we again remind you of the surface of a sphere: it's finite in area, but if you walk around on one you never fall over an edge or reach an end. For a two-dimensional being living on the sphere, there is nothing but that sphere. By analogy, it is possible that the topology of the Universe is such that it is finite, but you never reach an end and there's nothing surrounding it.

Time, too, could be infinite. There may be an infinite future ahead of us, and perhaps there is also an infinite past behind us. When physicists say that the Universe started with the Big Bang around 13.7 billion years ago, they really only mean that such a big bang happened, and created the Universe as we know it, but not necessarily that there was nothing before that.

These two possible infinities, of time and space, are potential in Aristotle's sense: even if they exist, you can never reach them. Aristotle was quite happy with this kind of infinity; they have no teeth. But he also mused about another type of infinity: an *actual* one that can exist right there in front of you, not at the end of some boundless process. Do actual infinities exist in nature? Are there quantities that you can measure that can become infinite?

One place where actual infinities might occur is at the centre of black holes. Black holes form when something very massive, like a star, collapses in on itself and becomes denser and denser.

Eventually a black hole forms, surrounded by a horizon. The horizon is the boundary of no return, if you cross that horizon you will never be able to get out again. As you move towards its centre, things get denser and denser, and theory suggests there is infinite density at its centre.

From the outside, however, you can't see that infinity. The horizon shields it from your view. This idea has led the physicist Roger Penrose (creator of Penrose tiles; see Chapter 5) to come up with a theory called *cosmic censorship*. The idea is that whenever an actual infinity, a 'naked singularity', forms in nature, it will always be hidden behind such a horizon – you will never see it from the outside. It's only a conjecture; no one has been able to prove that it is always true, but it's interesting. Even though mathematics gives us the power to imagine infinity, it may be that nature has taken care to hide it from us. If that's the case, then real, physical infinity will forever remain an unknown, a question mark, an x . . .

x Marks the spot

Xis a glamorous letter. It stands for the unknown, the mysterious, the forbidden, for kisses and for the special spot where a treasure is hidden. In maths it strikes fear in the heart of many: x stands for algebra and to many people that symbolizes the scariest aspect of maths – incomprehensible equations that have no connection with everyday life or common sense.

Yet, x makes things easier. No, really, it does! People have been trying to work out unknown quantities since the beginning of recorded history. How big should your field be to grow 300 bushels of grain? How many sarcophagi can you fit inside a pyramid? How much does one pint of beer cost if three cost £9? The Egyptians, who were very fond of problems involving beer as well as those involving pyramids, had a very cute word to describe an unknown quantity: 'aha'. Thus, they might be looking for the value of 1 aha if, as in our beer example, 3 ahas have the value 9. You can imagine an Egyptian beer drinker uttering a satisfied 'aha' when hitting on the answer, 3, but that wasn't the reason why the Egyptians chose that word. It rather boringly translates as 'heap'. A heap of money, or beer, or grain, or sarcophagi.

It was the scribe Ahmes who used the word 'aha' in the 18-feet-long Rhind papyrus, which he produced in around 1650 BC. Using a special word to describe a general unknown quantity is a clever idea. After all, it doesn't matter if you are thinking in terms of beer, grain or pyramids. The problem is to find the value of something, if

3 times the something is 9. Thinking in general terms and learning how to work out the solution, no matter what you are dealing with, gives you a recipe that works in many situations. It's a step towards abstraction.

Save your breath

Yet the Egyptians were still far away from what we would today recognize as algebra: something that involves only letters and symbols that stand for numbers and operations such as adding or subtracting. The Egyptians, like the Babylonians and many other cultures after them, talked about their maths problems using words. It's a convoluted business. 'Find the value of aha if 3 ahas have the value 9' is already quite hard to say, but in terms of modern algebra this translates to the neat equation

$$3x = 9.$$

This expression is not only short, but also gives you a visual aid to solving the problem. An equation is like an old-fashioned scale people used to use on markets to measure out quantities of apples, or grain, perhaps even casks of beer, though probably not sarcophagi. Whatever you do to one side, you need to do to the other; otherwise the scale tips out of balance. In our equation, to find out the value of x, you need to get a single x on one side, which you do by dividing the left-hand side by 3. To keep things equal you need to do the same to the right-hand side, giving the solution

$$x = 3.$$

The advantages of algebra are more obvious with a more complicated equation. Try describing the equation

$$3(x + 1) + 1 = 10$$

in words without simply reading off the algebra. Not only will you soon run out of breath, you will also have lost your visual aid that, with a bit of practice, immediately suggests that you should start by subtracting 1 from each side to get

$$3(x + 1) = 9.$$

The next step is to divide each side by 3, giving

$$x + 1 = 3,$$

and finally you subtract 1 again from each side to get

$$x = 2.$$

If you are trying to calculate areas or volumes, your equations will have to include things like x^2 and x^3, and you can imagine how complicated the word version of these equations becomes.

Stubborn words

Given how handy the use of letters and symbols is, it's surprising how long it was in the making. It's even more surprising that the development of algebra and the development of symbolic maths aren't one and the same thing. The word 'algebra' doesn't come from the Greek, or Latin, but from the Arabic. It derives from the name of a book, the *al-jabr wa'l-muqabala,* written by Abdallāh Muhammad ibn Mūsā al-Khwārizmī, a scholar who worked at Baghdad's famous House of Wisdom round about AD 800. Other mathematicians before him, for example the Greek Diophantus in around AD 250, and Brahmagupta, the foresighted inventor of zero, whom we met in Chapter 0, had used some symbolic maths in their algebra. But al-Khwārizmī was having none of that: he used words only, not a symbol in sight. Yet he is today known as the true father of algebra, for giving exact methods for finding solutions to equations. His methodical nature is remembered in the word 'algorithm' that we use today to describe a formulaic recipe, something a computer can use, to solve a problem. It comes from the Latin translation of al-Khwārizmī's name, *Algoritmi.*

Although algebra kept on developing, symbols didn't really take hold in maths until the 15th century. From then on, European mathematicians from all countries and backgrounds painstakingly invented them, one by one. The symbols we use for plus and minus, + and –, were introduced in 1489 by the German Johann Widman, who was busy facilitating the maths needed to keep warehouses in order. Another German, Christoff Rudolff, working a little later, was responsible for our symbol for square roots. The equals sign was introduced by the Englishman Robert Recorde in 1557. The multiplication sign first appeared in a publication from 1631 by another Englishman, William Oughtred. And the division sign, ÷, is credited to the Swiss mathematician Johann Rahn, making its first appearance in 1659.

x, y, z

It was a Frenchman, however, who gave the letter *x*, and its cousins *y* and *z*, their special status. In his work *La géometrie*, published in 1637, René Descartes used letters from the end of the alphabet for unknown quantities. And he used letters from the beginning of the alphabet to stand for fixed, but unspecified, numbers. It's a practice we continue today. Our beer problem above can be phrased in even more general terms, by writing *a* and *b* for the numbers 3 and 9. The equation now becomes

$$ax = b,$$

and its solution is

$$x = \frac{b}{a}$$ (assuming that *a* isn't equal to 0).

This general solution will work no matter what the *x* stands for, and it will work whenever some number *a* times *x* equals another number *b*. Simply plug in your specific values for *a* and *b* and, bingo, the result pops out.

It's this ability to express things neatly and concisely in most general terms that makes symbolic mathematics so powerful. In effect, the strange symbols and abstract looking equations that frighten so many people are nothing more than a language that enables us to express things that are too hard to pin down in words. It's a language that takes some getting used to, but it shouldn't scare you more than French, Spanish or Greek. Once you have learnt some vocabulary and the glue that holds it together, logic, you can start understanding some of its sentences and even make your own.

Descartes' is probably the first work of maths that resembles what you see in maths books today. Apart from letters it uses many familiar symbols, and it expresses taking powers of variables in the same way as we do today, using a little number written as superscript (as in x^2). But Descartes is much more famous for other contributions to mathematics, science and philosophy. 'I think, therefore I am,' is his famous answer to excessive scepticism. Even if you can't be sure that all those things you see around you – this book, the table, your cup of tea – really exist, you can be sure that you do, because you are thinking. Descartes described this line of thought in his philosophical treatise *Discours sur la méthode* which, tucked in at the end, also contained the first ever mathematical explanation of why a rainbow has the shape it does. In mathematics his name lives on in something we're so used to today that we hardly notice that it needed to be invented at all: the Cartesian coordinate system.

x marks the spot

Descartes liked to stay in bed until noon, and according to a (probably untrue) story he came up with his coordinate system while engaged in that noble activity. He observed a fly on the ceiling and wondered how to describe its location. One way of doing this is to pick one of the four corners of the room – say the bottom left corner of the ceiling as it appears in your visual field – and to specify the distance you have to move to the right, and the distance you have to move up to get to the fly, starting off at your chosen corner. The location of the fly, as long as it's staying put, is therefore given by two numbers, its coordinates, reflecting the fact that the ceiling it's moving on is two-dimensional.

We have already met this way of describing the location of points – Cartesian coordinates – in Chapter 4. We can extend it by imagining that the ceiling continues indefinitely in all directions and taking two perpendicular axes to replace the left and bottom boundaries of the ceiling. Each point has two coordinates: the first gives its location in the horizontal direction (with negative numbers meaning it is to the left of the vertical axis), and the second gives its location in the vertical direction (with negative numbers meaning it is below the horizontal axis).

Cartesian coordinates are everywhere in everyday life. Just think of the last time you saw a graph in a newspaper, visualizing, say, the fluctuation of share prices over time. It comes from marking time on a horizontal axis and share price on a vertical axis. If on day x the share price was y, you mark the point with coordinates (x, y) on the coordinate system given by your two axes. The jagged curve you see comes from doing this for all days x in the time period you are interested in. There are many, many other situations in life in which two variables are related to each other, and whenever that's the case Cartesian coordinates help you visualize that relationship.

From symbols to shapes

In mathematics Cartesian coordinates had a huge impact, because they blurred the boundaries between algebra and geometry in an unprecedented way. As we saw in Chapter 4, you can use these coordinates to describe geometric shapes. A circle of radius r around the point $(0, 0)$ consists of all points whose coordinates (x, y) satisfy the algebraic equation

$$x^2 + y^2 = r^2.$$

An even simpler shape appears when you consider all the points whose two coordinates x and y are equal, which satisfy the equation $y = x$, such as (0, 0), (1, 1), (2, 2), (3, 3) and so on. It gives you the straight line that passes through the point (0, 0) and exactly halves the right angle between the axes.

You can make the slope of the line steeper by multiplying the x in the equation by a number greater than 1, for example

$$y = 2x,$$

or less steep by multiplying it by a number less than 1, for example

$$y = \frac{1}{2}x.$$

And if you multiply x by a negative number, as in

$$y = -2x,$$

your slope switches around from running upward as you go from left to right, to running downwards.

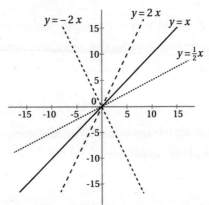

If you'd like your line to lie further up or down on the coordinate grid you simply add or subtract a number. For example, the equation

$$y = 2x + 1$$

meets the vertical axis 1 unit above the point (0, 0) and the equation

$$y = 2x - 1$$

meets the vertical axis 1 unit below it.

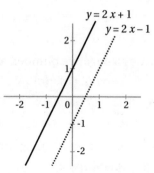

The great thing is that you can get *all* straight lines in this way. This gives you a general form for the equation of a straight line:

$$y = ax + b.$$

The numbers *a* and *b* determine the slope of the line and its location in the plane respectively. A point with coordinates (*x*, *y*) lies on the line if *x* and *y* satisfy the equation.

Algeometry or geomalgebra?

This way of representing things enables you to approach geometric problems using algebra and algebraic problems using geometry. For example, imagine you are asked to find all the numbers x and y that satisfy the two equations

$$y = -x + 6$$

and

$$y = 2x.$$

If you really don't like doing algebra, you could do this by recognizing that both equations represent straight lines. By plugging the numbers 0 and 1 in for x in the first equation you see that the first line passes through the points (0, 6) and (1, 5). Doing the same for the second equation you see that the second line passes through (0, 0) and (1, 2). Now it's easy to plot those two lines in Cartesian coordinates because all you need to do is connect those pairs of points using a ruler. If a pair of numbers x and y satisfy both equations, then the corresponding point (x, y) lies on both lines. Therefore the solution to your algebraic problem is given by the point at which the lines intersect. Looking at your plot you see that this point is given by the coordinates (2, 4): the numbers $x = 2$ and $y = 4$ satisfy your two equations.

Conversely, imagine you are asked to solve the geometric problem of finding the intersection of the two lines, but you haven't got a ruler, pencil and paper. If you're a little versed in algebra, you could simply solve the two equations

$$y = -x + 6$$

and

$$y = 2x,$$

together and thereby solve your problem algebraically.

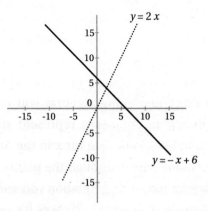

The point at which the lines meet, (2, 4), gives the numbers that satisfy both equations.

This problem involving two straight lines is fairly basic, of course, but the point is that *any* algebraic relationship between two variables *x* and *y* can be represented by a geometric shape drawn in Cartesian coordinates: it consists of all points whose coordinates (*x*, *y*) satisfy the relationship. Conversely, many shapes we are familiar with from geometry can be described algebraically. For example, each of the conic sections we met in Chapter 60, so beloved by the Greeks, comes with its own equation, characterized by the fact that the power of *x* in those equations is 2 (see the box).

The gently undulating sine and cosine waves we met in Chapter τ are the curves you get from the equations

$$y = \sin(x)$$

and

$$y = \cos(x).$$

Today many students of maths think of those geometric shapes and their algebraic representation as being one and the same thing – it's a testament to the power of Descartes' ideas and those of other contemporaries who also contributed to the algebraification of geometry. The seeds they sowed were instrumental in the development of calculus, which is all about finding the rates of change you get when you vary two dependent variables (see Chapter *e*). Today those ideas have blossomed into a whole mathematical area of its own, called *algebraic geometry*. One very notable contemporary of Descartes, who independently invented the Cartesian coordinate system, was Pierre de Fermat, whose famous last theorem we have already met in Chapter $\sqrt{2}$. In a very pleasing development it was tools from algebraic geometry that eventually helped to prove Fermat's last theorem, over 350 years after he first posed it.

The price of laziness

How many of Descartes' ideas were conceived in bed we don't know, but it seems that his fondness for staying in it all morning may have led to his untimely death in 1650, at the age of only 54. He

CONIC SECTION EQUATIONS

The equation

$$(ax)^2 + (by)^2 = 1$$

represents an ellipse. When a and b are equal, this is just the equation for a circle.

The equation

$$(ax)^2 - (by)^2 = 1$$

represents a hyperbola. The equation

$$y = ax^2$$

represents a parabola.

was in Sweden at the time, having been asked by Queen Christina to be her maths tutor. Unfortunately the Queen preferred to work early in the morning. Some say that these early hours and the Scandinavian temperatures caused the pneumonia that eventually killed Descartes. Others have suggested that he was poisoned by a Catholic priest worried about Descartes' radical theology. Either way, our little letter x owes Descartes a very large drink for the special status he bestowed on it.

Hopefully we have convinced you that x isn't scary or weird, but just a friendly word in the language of mathematics. Usually it is just a placeholder for something real and harmless like a number. But its role as symbol for the unknown leads to a much stranger and exotic idea: imaginary numbers.

i Imaginary is everything

Sometimes, maths trips you up. You set off answering what seems a perfectly innocuous question, one you have answered many times before. You head off, applying your tried and tested methods, but quickly hit a dense fog of confusion. You look down to find you have stumbled off the very edge of the number line, finding yourself in a strange and unknown mathematical landscape.

When you find yourself teetering on the edge of the unknown, you have several choices. You could close your eyes, slowly back away and pretend nothing has happened. Or you could hold your breath, gingerly making your way through the unknown territory, until you are back on solid mathematical ground. Alternatively, you could do what the best adventurers do: explore the strange new world, seeking out new mathematics, boldly going where no mathematician has gone before.

Close your eyes and back away

Heron of Alexandria found himself in such a situation at the end of the first century AD. In his book *Stereometria*, he published a neat formula to calculate the height and volume of a *frustrum* – a truncated pyramid with a square base and top, joined by straight sloping sides. He demonstrated his formula first with the example of a frustrum with a square base with sides of length 10, top with sides of length 2 and sloping sides of length 9. He could easily calculate that the height, h, of such a shape was 7, using his formula:

$$h = \sqrt{c^2 - \frac{(a-b)^2}{2}} = \sqrt{81 - \frac{64}{2}} = \sqrt{49} = 7,$$

where a and b are the lengths of the sides of the bottom and top squares and c is the length of the sloping edge.

Heron then tried this approach for a different example, with a base of side length $a = 28$, top of side length $b = 4$ and sloped edge of length $c = 15$. This time his formula gives the height as:

$$h = \sqrt{c^2 - \frac{(a-b)^2}{2}} = \sqrt{225 - 288} = \sqrt{-63}.$$

But, hang on a second: the answer to the problem, the square root of –63, is the number that when multiplied by itself gives you –63. That's how square roots work. But, as we saw in Chapter 1, any number multiplied by itself gives you a positive number – either the original number is positive, and positive times positive is positive, or it is negative, and negative times negative is also positive. So the number $\sqrt{-63}$ doesn't exist.

Heron had managed to come up with an impossible example – no frustrum with these measurements could exist. And he had come up with the first recorded example we have of a calculation involving the square root of a negative number. The answer was recorded in his book, however, as $\sqrt{63}$. We don't know if he deliberately set his jaw and ignored the strange result or simply made a mistake and didn't notice. But it seems that when offered a tantalizing glimpse into unknown mathematical lands, he closed his eyes and backed away.

Mistakes and mathematical glory

We can easily forgive Heron for overlooking his mathematical discovery. We all (some of us more than others) often make silly errors in our working, so we are inclined to think we have just messed up our arithmetic or algebra rather than that we have missed a world-changing discovery. It took many centuries for someone to realize that this wasn't just a mistake, and that instead Heron had glimpsed a strange new mathematical territory.

Zooming forward to the sixteenth century we find a time when algebra was thriving and mathematicians challenged each other to solve equations, not just to boost their ego, but also to enhance their reputations, advance up the career ladder and even to win prizes. One favourite challenge was to find the solutions of *cubic equations*. These were equations involving some unknown quantity, say x, where the highest power of x is 3, for example,

$$x^3 - 6x - 4 = 0.$$

Any cubic equation can be written as

$$x^3 + ax^2 + bx + c = 0,$$

for some values of a, b and c.

Mathematicians had been puzzling over cubic equations since the ancient Greeks and Egyptians, partly because cubic equations can arise from physical or geometric problems. For example, the volume of a cube or sphere is given in terms of the cube of its side length or radius.

What they really wanted was a cubic equivalent to the quadratic formula that we saw in Chapter 12 that gave all the solutions to any quadratic equation (those with the highest power of x being 2). What's more, they knew that every cubic equation had at least one value of x that worked – that is, there was always some number that, when plugged in for x, would make the expression equal to 0.

It's not too hard to see why, using our example

$$x^3 - 6x - 4 = 0.$$

If you plug in a very large positive number for x, say $x = 1,000$, then this will swamp all the other terms and make the left-hand side of your equation very large and positive:

$$1,000^3 - 6 \times 1,000 - 4 = 999,993,996.$$

Equally if you plug in an x that is very large and negative, again this term will swamp all the others but this time will make the left-hand side of your equation very large and negative:

$$(-1,000)^3 - 6 \times (-1,000) - 4 = -999,994,004.$$

As you slide the number x up the number line from a very large negative number to a very large positive one, the left-hand side of the equation also varies continuously – with no breaks or sudden leaps – from a very large negative to a very large positive number. At some point on this journey it must pass through 0, so there must be some value of x that gives the value of 0. That value is a solution to the equation.

A mathematical battle

In the 16th century many tricks were known for finding the values of x that satisfied particular types of cubic equations but a general solution remained elusive. As mathematicians discovered these new techniques they often kept their discoveries secret to give themselves an advantage in mathematical contests. This meant that great mathematicians such as Scipio del Ferro of Bologna, Italy, published little, but we know that he was the first person to make great headway in the battle against the cubic. He discovered a general formula for solving *depressed cubics*: those cubic equations that do not contain any squared powers. At the time these were termed 'unknown and cubes equal to numbers', so something of the form $x^3 + bx = c$.

Another mathematician who had success with cubics was Niccoló Fontana, who was more commonly known by his nickname, Tartaglia. He was a brilliant young mathematician, who overcame a tragic childhood. Not only was he born into a very poor family, his father was murdered when he was six years old, plunging his family into total poverty, and he was nearly killed by soldiers when he was just 12. When the French army captured his home town, a soldier wounded the young Tartaglia with a sabre, cutting his jaw and palate. His mother nursed him back from the brink of death but he was left with disfiguring scars and a stammer, leading to his nickname of Tartaglia, or 'stammerer'. He went on to teach himself mathematics, eventually finding a patron who enabled him to study and become a mathematics teacher. Tartaglia found a way to solve another type of cubic, of the form $x^3 + ax^2 = c$, or 'squares and cubes equal to numbers' as they were described at the time.

On his death bed in 1526, del Ferro passed on his knowledge to his student, Antonio Fiore. Once Fiore had this ace up his sleeve, he challenged Tartaglia to a public mathematical duel. Tartaglia posed a variety of questions for Fiore to solve, while Fiore, a more mediocre mathematician, relied on his trump card and set Tartaglia thirty different versions of depressed cubics. Fiore's bid for glory backfired as Tartaglia succeeded in independently discovering the same method for depressed cubics just in the nick of time, the night before the contest: Tartaglia 1, Fiore 0.

Fortunately for us, the great Italian mathematician Giralomo Cardano was privy to such treasures. Tartaglia finally divulged his secrets to Cardano on condition that he 'swore a most solemn oath, by the Sacred Gospels and his word as a gentleman, never to publish the method'. However, when Cardano became aware of del Ferro's earlier discovery of the same method he included it in his great work, *Ars Magna* (meaning 'The Great Art') in 1545.

Cardano justified his actions by saying he had published del Ferro's method, rather than breaking his oath not to publish Tartaglia's, and gave both mathematicians due credit. But Tartaglia was enraged. A terrible feud ensued, which resulted in a final epic battle in a church in Milan on 10 August 1548, before a large audience. In the blue corner was Lodovico Ferrari, Cardano's student, who had worked with him on *Ars Magna* and was fighting on his behalf; and in the red corner was Tartaglia. When the final bell rang (in this case, probably the dinner bell, as Tartaglia said the meeting finally broke up as supper time drew near) it appears Ferrari was the one with his fist in the air. He went on to have a distinguished mathematical career, riding the wave of mathematical glory resulting from their contest, while Tartaglia lost his university position. Proof that these competitions had very tangible consequences for the winners and losers – all's fair in love and maths.

Hold your breath and hope for the best

Del Ferro and Tartaglia had independently come up with a formula for solving any depressed cubic equation. In *Ars Magna*, Cardano showed that any cubic equation can be written as a depressed cubic using a tricky substitution of variables. Therefore he could then use del Ferro and Tartaglia's formula to solve any cubic equation.

But Cardano's brilliant insights went even further than solving a centuries'-old problem. The depressed cubic formula sometimes generated solutions involving the square root of negative numbers. For example, applying the formula to our example $x^3 - 6x - 4 = 0$ gives the solution as

$$x = \sqrt[3]{(2 + 2\sqrt{-1})} + \sqrt[3]{(2 - 2\sqrt{-1})}.$$

(The symbol $\sqrt[3]{n}$ stands for the *cubic root* of n, which, as you'd expect, is the number that when cubed gives you n. For example $\sqrt[3]{8} = 2$.)

Unlike previous mathematicians – Heron centuries earlier, and his peers del Ferro and Tartaglia – Cardano did not ignore these strange solutions or pretend that they weren't there. He might not have liked them, but he gritted his teeth and became the first person to explicitly calculate with the square root of negative numbers, treating them like any other number.

The first recorded example of this is in *Ars Magna* where Cardano asks for two numbers that add to 10 and multiply to give 40. Using known techniques for solving the corresponding equations (see the box overleaf) he found that the solution was

$$x = 5 + \sqrt{-15}$$

and

$$y = 5 - \sqrt{-15},$$

which involved the illegal square roots of negative numbers. Here Cardano held his breath, put 'aside the mental tortures involved' and carried on. If you multiply $5 + \sqrt{(-15)}$ by $5 - \sqrt{(-15)}$, as if they were any other numbers, you get

$$(5 + \sqrt{-15}) \times (5 - \sqrt{-15})$$

$$= 25 - 5 \times \sqrt{-15} + 5 \times \sqrt{-15} - \sqrt{-15} \times \sqrt{-15}$$

$$= 25 - (-15)$$

$$= 25 + 15$$

$$= 40,$$

a perfectly legal number.

Cardano was the first person to step off the number line into the unknown, a revolutionary step for mathematics. But he was only happy to linger there briefly, as long as the way back to the *real numbers* was clear. And he was in good company. Although the influential mathematician Descartes, who gave us the Cartesian coordinates we use to describe the position of any point on the plane or higher-dimensional space (see Chapters 4 and x), wrote about square roots of negative numbers in his famous work *La geometrie*, he was just as queasy about them as Cardano. Descartes dismissed them as 'imaginary' as they represented solutions that could not be physically or, to him, geometrically imagined, thus coining their name.

CARDANO'S PROBLEM

Cardano wanted to find two numbers that add to 10 and multiply to 40. Writing x and y for these numbers, this translates to

$$x + y = 10,$$

$$xy = 40.$$

From the first equation we get that

$$y = 10 - x.$$

Plugging this in for y in the second equation gives

$$x(10 - x) = 40.$$

This can be arranged to give the quadratic equation

$$x^2 - 10x + 40 = 0.$$

It is this equation that Cardano solved to get the solutions

$$5 + \sqrt{-15}$$

and

$$5 - \sqrt{-15}.$$

To boldly go . . .

One person who was not as easily scared by those 'imaginary' numbers was the great Leonhárd Euler. We've already met Euler several times in our travels. That's because he remains one of the most prolific mathematicians who ever lived and who made contributions to many areas of maths, many of which he invented. Euler carried on with his mathematics even when he became totally blind from cataracts in his 60s. This included writing the textbook *Elements of Algebra*, which he dictated to a young servant, who no doubt learnt a great deal of maths along the way.

In his famous book *The Elements*, Euler said that square roots of negative numbers 'are usually called imaginary quantities, because they exist merely in the imagination . . . notwithstanding this . . . nothing prevents us from making use of these imaginary numbers, and employing them in calculation'. It was thanks to Euler that we use the letter i for $\sqrt{(-1)}$. He was entirely happy using what we now call *complex numbers*, numbers that combine ordinary and imaginary numbers, such as

$$1 + 2\sqrt{-1} = 1 + 2i$$

or

$$10 - 5\sqrt{-1} = 10 - 5i.$$

In general a complex number is a number that can be written as $a + ib$, where a and b are ordinary real numbers. The first part, a, which doesn't involve the strange number i, is appropriately called the real part of the complex number and the second part b is called the imaginary part.

Unlike Cardano and his 16th-century colleagues, who gingerly stepped into complex numbers as a quick detour to get them back onto the ordinary number line, Euler was happy to wallow about in the complex mire, like the proverbial pig in mud. Once he'd entered the new realm he set off to explore it and see what he could find. Using the many ways in which you can write the number e (see Chapter e), he worked out what it means to raise e to the power of an imaginary number. And he discovered one of his most famous results that led to one of the best-loved equations in mathematics:

$$e^{i\pi} + 1 = 0.$$

Euler's identity, as it is known, is thought by many mathematicians to embody mathematical beauty. It gives an elegant, and deceptively simple, relationship between five of the most important numbers in mathematics (0, 1, e, π and i) using four of the most important mathematical operations and relations (addition, multiplication, exponentiation and equality).

Euler's identity is a special case of a more general result, *Euler's formula*:

$$e^{i\theta} = \cos \theta + i \sin \theta.$$

This involves the trigonometric notions of sine and cosine, which we met in Chapter τ. This remarkable result gave mathematicians a way to extend the exponential function to any number, complex or real. Also, unknown to Euler, this formula gives us two different ways to finally visualize complex numbers, allowing them to leap from our imagination onto the page before our very eyes. But this great leap was made not by Euler, one of the greatest mathematicians of all time, but instead by a humble Norwegian surveyor by the name of Caspar Wessel.

Mapping uncharted territory

Wessel was a surveyor and mapmaker for the Danish government and he developed many sophisticated mathematical techniques to produce accurate maps using triangulation and trigonometry. He often appended articles to his geographical work to explain his new theoretical ideas. This innovative approach led him to write a paper that was purely on mathematics. On 10 March 1797, Wessel's paper 'On the analytic representation of direction' was presented to Royal Danish Academy of Sciences. Wessel couldn't present his paper personally as he wasn't a member of the Academy. Although his paper was accepted, the first to be accepted from a non-member, and was published in their journal two years later, it never had the impact it should have had on the mathematical community. Instead his revolutionary ideas were made famous when they were rediscovered a few years later by the amateur French mathematician Jean-Robert Argand.

Wessel's revolutionary idea, probably inspired by his work as a surveyor and the mathematical techniques he developed there, was to visualize the complex numbers as an extension of the ordinary number line to a plane, called the *complex plane*.

A complex number such as $3 + 4i$ is specified by a pair of numbers, in this case (3, 4). Such a pair, as we saw in Chapter x, specifies the coordinates of a point on the plane with two perpendicular axes: it's the point you get to by taking 3 steps in the horizontal direction and then 4 steps up in the vertical direction. Generally, any complex number $a + bi$ can be represented by the point with coordinates (a, b) in the Cartesian coordinate system we know and love.

This neat representation also tells us what it means to add or subtract two complex numbers, say $1 + 2i$ and $3 + 4i$. To add or subtract them, you separately add or subtract their real and imaginary parts:

$$(1 + 2i) + (3 + 4i) = (1 + 3) + (2 + 4)i = 4 + 6i,$$

$$(1 + 2i) - (3 + 4i) = (1 - 3) + (2 - 4)i = -2 + -2i.$$

Representing those numbers as the points (1, 2) and (3, 4) on the plane, adding them simply corresponds to adding the points' coordinates: their sum corresponds to the point with coordinates (4, 6) and their difference to the point with coordinates (−2, −2). Easy!

The right direction

Wessel thought of complex numbers $a + ib$ as the point (a, b) on the plane, but more importantly for him, given his background, they also represented a direction and length. A point (a, b) in the plane can also be specified by drawing an arrow from the point where the axes cross (the point (0, 0), the origin) to the point in question, (a, b).

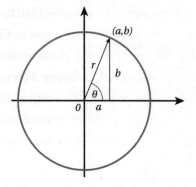

The unit circle

That arrow is specified by its length, call it r, and the angle, call it θ, the arrow makes with the horizontal axis (see the diagram). Using Pythagoras' famous theorem and some trigonometry you can relate the Cartesian coordinates (a, b) of the point to the length r and angle θ of the arrow it defines (see the box overleaf). The relationship is

$$r \sin \theta = b$$

and

$$r \cos \theta = a.$$

This means that you can write a complex number $a + ib$ equally well as

$$r \cos \theta + i \times (r \sin \theta)$$
$$= r (\cos \theta + i \sin \theta).$$

WRITING COORDINATES IN TERMS OF ARROWS

Suppose you have a point with Cartesian coordinates (a, b) and you want to specify the arrow that starts at the point $(0, 0)$ and points to (a, b). Pythagoras' famous theorem (see Chapter $\sqrt{2}$) tells us that the length of that arrow is

$$r = \sqrt{(a^2 + b^2)}.$$

Trigonometry tells us that the angle θ the arrow makes with the part of the horizontal axis to the right of $(0, 0)$ is given by the relationships

$$\sin \theta = \frac{b}{r}$$

and

$$\cos \theta = \frac{a}{r}.$$

This means that

$$r \times \sin \theta = b$$

and

$$r \times \cos \theta = a.$$

And by Euler's formula we know that

$$r (\cos \theta + i \sin \theta) = re^{i\theta}.$$

Writing a complex number in the form $a + ib$ immediately tells us the Cartesian coordinates of the point in the plane it represents, while writing it in the form $re^{i\theta}$ specifies the length and angle of the arrow to that point. When you see the slightly frightening expression $re^{i\theta}$, forget about the numbers e and i and what they represent, and think in terms of an arrow with a given length and angle.

With Wessel's new understanding, Euler's famous identity is suddenly obvious: $e^{i\theta}$ is the arrow of length 1 that is at an angle of θ to the positive x-axis. Since π, as an angle measured in radians, is equivalent to 180 degrees, the arrow $e^{i\pi}$ lies along the horizontal axis pointing to the left, ending at the point $(-1, 0)$. On the complex plane it specifies the number

$$-1 + 0 \times i = -1,$$

so we can immediately see that

$$e^{i\pi} = -1$$

$$e^{i\pi} + 1 = 0.$$

But Wessel's revolutionary insight also made sense of what it means to multiply complex numbers. Suppose the first complex number is given by an arrow of length r that makes an angle θ with the horizontal axis and the second number is given by an arrow of length s that makes an angle φ with the horizontal axis. This means that our numbers can be written as $re^{i\theta}$ and $se^{i\varphi}$, so to multiply them we need to work out

$$re^{i\theta} \times se^{i\varphi}.$$

The rules of exponentiation (see Chapter e) tell us that to do this we multiply r and s separately and then simply add the exponents of e:

$$re^{i\theta} \times se^{i\varphi} = rs\, e^{i(\theta + \varphi)}.$$

This gives us a new complex number, specified by an arrow of length rs and an angle $(\theta + \varphi)$. So multiplying $re^{i\theta}$ by $se^{i\varphi}$ amounts to first scaling the arrow that defines $re^{i\theta}$ (multiplying it by s to make it longer or shorter) and then rotating (changing its angle by adding φ to it). The multiplication of those strange abstract beasts has suddenly become tangible.

Wessel's desire for a simple way to represent and manipulate directions gave us a tangible way to picture complex numbers and how they interact. What had once seemed imaginary, even terrifying, was now crystal clear.

Once complex numbers had been tamed, it quickly became apparent how useful they were. They made describing rotations in the plane a simple operation of multiplication: for example if you want to rotate a vector by 90 degrees, multiply it by i. When the Irish mathematician William Rowan Hamilton learnt this effortless description of rotations in two dimensions, he set off on a thirty-year quest to provide a similar description of rotations in three dimensions. He finally discovered the answer, in a flash of genius, while walking under the Broome Bridge in Dublin (the moment is commemorated with a plaque under the bridge). His extension of the complex numbers to three dimensions produced numbers called *quarternions*.

And now, to the movies . . .

You come across Hamilton's ingenious discovery whenever you go to the movies, or at least when the movie you are watching contains a good amount of computer-generated imagery. Imaginary creatures, such as the loveable Gollum from the *Lord of the Rings* and *The Hobbit,* owe their existence to a very generous helping of mathematics. When these beasts are created, animators often start by representing them as a wire mesh made up of lots of little individual facets, often triangles, as we saw in Chapter 3. This mesh can easily be captured in a computer, by representing the corners of the triangles by their coordinates in the three-dimensional extension of Descartes' coordinate system.

To bring the wire skeleton to life you need to colour in the individual facets of the mesh. If you want your creature to look realistic, the colouring needs to take account of the lighting of the scene it's in. Maths is indispensable here because, as we also saw in Chapter 3, you need to calculate whether a light ray from, say, the Sun, or a lamp, would hit a particular facet – if it does, you need to colour that facet brighter.

But the exciting point comes when you make your creature move. Capturing the movements of real people is an important part of the process: you film people moving around with reflectors strapped to pivotal parts of their bodies and then feed the changing positions of those reflectors into your computer, to help you make your creature move realistically. But you also need fast and efficient techniques to rotate things, perhaps part of an arm or leg, or the spinning of an object. These objects are represented by the coordinates of the surface facets. And this is where complex numbers and, in particular, quaternions come in. Those inventions, which have their roots in the race to solve the cubic equation, provide the tools to rotate.

This is our favourite application of complex numbers, but by no means the only one. One of their strong points is that they capture two pieces of information – two coordinates, or the length and angle of an arrow – in one number. They are vital in electrical engineering, in calculating the voltage and currents in circuits. They are also used to make life easier in fluid dynamics, signals processing and many other areas of science. You can be sure that something you come across every day, whether watching the latest blockbuster or taking a call on your mobile phone, is possible thanks to that step into the imaginary world.

QED

W e've come to the end of our journey through mathematics and hope you have enjoyed our favourite mathematical sights and stories. We've picked them from the many ideas mathematicians have shared with us over the years. They've chatted to us over coffee or wine, given their time for in-depth interviews and guided us through the dense thicket of difficult mathematical papers. We are always grateful for their patience, generosity and enthusiasm.

Marianne was helped on her fledgling journey into maths by Shaun Bullett, Professor of Mathematics at Queen Mary, University of London, who patiently supervised her PhD and guided her first steps into the world of mathematical research. Rachel started on her path with the help of Bob Sullivan, who first encouraged her passion for exposition, and Cheryl Praeger, who first inspired her interest in pure maths in the beautiful surrounds of the University of Western Australia. And, of course, her time working as a consultant for John Henstridge at Data Analysis Australia opened her eyes to the impact and power of mathematics in all aspects of life.

Thanks to Quercus for suggesting such an interesting project and also to our lovely agent Peter Tallack for making it happen. In particular we'd like to thank all the people at Quercus, including the proofreaders, copy editors and illustrators for their help and support in bringing this book to reality.

Helen Joyce, now editor of the International Section of *The Economist*, first revealed the world of maths communication to us. As the previous editor at *Plus* Magazine, Helen shared both her passion for maths and her expertise at telling mathematical stories to a wider audience. Marcus du Sautoy, Simonyi Professor for the Public Understanding of Science and Professor of Mathematics at the University of Oxford, taught Rachel a great deal about storytelling when they went on a mathematical tour of the city. And we have both benefitted from the years of experience – both mathematical and authorial – of John D. Barrow, Professor of Mathematical Sciences at Cambridge and director of the Millennium Mathematics Project that *Plus* is part of. As our office neighbour at Cambridge, John not only provides a running commentary on various sports (white noise to us), but also imparts his deep insights into many areas of maths and his skill at communicating them, proven in his many popular science books.

And of course, this book would not have been possible without the support and encouragement of our friends (mathematical and maths-phobic) and family. Rachel would like to thank her parents for their curiosity, Charles for his belief in her, and Henry and Elliott for reminding her of the excitement of discovering both mathematics and writing. Marianne would like to thank her parents for all their love and support, and for planting the seed of her love of maths, and all her friends for their patient support through those long weekends of book writing.

It's the end of the book but it's by no means the end of the journey. One of the reasons we wanted to write this book was to dispel the myth that maths is 'done and dusted' and reveal what it really is – alive, dynamic and vitally important to the world we live in. Its role in technology, in understanding the complexities of nature and in guiding us to the furthest reaches of space and time, mean that maths is integral to our lives. If anything, it's going to play an increasingly important role, even if it often remains hidden below the surface. As Galileo noted in the 17th century, mathematics is the 'language of the Universe'. The most exciting question of all is: what will it tell us next? We can't wait to find out.

Get me some of that special source!

One of the nicest things about writing this book has been rereading some of our favourite books. We've also found it very exciting to read some of the original sources we mention in the book – such as reprints of ancient Indian textbooks, digital versions of Shannon's 1930s masters thesis, and exploring the delicate pages of a 19th-century copy of John Napier's *Wonderful Canon of Logarithms*. Here are some of our favourite sources that we've used in writing this book. We recommend you to take a look!

Chapter 0

The Book of Nothing, John D. Barrow, Pantheon, 2001. A thoroughly readable account of nothing by our good friend (and boss!) John Barrow.

Colebrooke's Translation of Bhaskara's Lilavati, Asian Educational Services, New Delhi, 1993. One of the most poetic textbooks we've ever read. It's much more fun doing maths problems when they involve bees, lotus flowers and crocodiles.

Chapter 1

To see the beginnings of our digital world, read Shannon's revolutionary paper 'A mathematical theory of communication',

The Bell System Technical Journal, Vol. 27, July 1948, pp. 379–423; October 1948, pp. 623–656: available at: cm.bell-labs.com/cm/ms/what/shannonday/shannon1948.pdf

You can step back even earlier, to Boole's wonderful book *An Investigation of the Laws of Thought, on which are Founded the Mathematical Theories of Logic and Probabilities*, published in 1853 and available at: www.gutenberg.org/files/15114/15114-pdf.pdf

Chapter √2

Chaos: Making a New science, James Gleick, Vintage Books, 1997. A very useful and informative account of the development of chaos theory.

Fermat's Last Theorem, Simon Singh, Fourth Estate, 2002. A very accessible and engaging story of the experience of hunting and finally solving one of the great problems in maths.

Chapter φ

The Golden Ratio: The Story of Phi, the World's Most Astonishing Number, Mario Livio, Broadway Books, 2002. A really enjoyable and comprehensive account of the wide reach of this number.

And there's a mine of information at Ron Knott's website, Fibonacci numbers and the golden section, available at: www.maths.surrey.ac.uk/hosted-sites/R.Knott/Fibonacci/fibnat.html

Chapter 2

The Music of the Primes: Why an Unsolved Problem in Mathematics Matters, Marcus du Sautoy, Harper Perennial, 2004. A fantastic journey through the primes and the Riemann Hypothesis.

Chapter *e*

The Construction of the Wonderful Canon of Logarithms; and their Relations to their Own Natural Numbers, John Napier, 1619, translated from Latin into English by William Rae Macdonald, 1888. We couldn't quite believe it when the version we ordered at the library was the actual 1888 copy of the translation, its yellowed pages tied together with a faded ribbon.

Chapter 6

Six Degrees: The Science of a Connected Age, Duncan J. Watts, Vintage Books, 2003. A great overview of a new discipline.

Sync: The Emerging Science of Spontaneous Order, Steven Strogatz, Penguin, 2004. A brilliant account of how and why research is done. Reading this book was one of the few times we've been tempted to return to mathematical research!

Linked: How Everything is Connected to Everything Else and what it Means for Business, Science and Everyday Life, Albert-László Barabási, Plume Books, 2003. A fascinating story linking up so many areas of our lives, brilliantly told.

Chapter 42

To find out what can happen when you simulate complex systems, we recommend that you read, watch or listen to Douglas Adams' classic *The Hitchhiker's Guide to the Galaxy*.

Chapter 43

In researching this book we came across one of the most entertainingly written papers we've read, by one of our favourite mathematicians, Freeman Dyson. 'Some guesses in the theory of partitions' appeared in *Eureka* (the annual journal published by the Archimedians, the Cambridge University Mathematical Society) in 1944. The last line alone makes it worth the read.
Available at: www.math.ucla.edu/~pak/papers/Dyson-Eureka.pdf

Chapter 60

Dava Sobel's wonderful book *Longitude: The True Story of a Lone Genius Who Solved the Greatest Scientific Problem of His Time* gives a fascinating account of the struggle and solution of the longitude problem.

Chapter 100

'The library of Babel', in *Fictions*, Jorge Luis Borges, translated by Andrew Hurley, Penguin, 1944.

For an excellent study of the impact of probability on real life, we strongly recommend you read, watch or listen to Douglas Adams'

The Hitchhiker's Guide to the Galaxy. And remember, in the face of improbability, DON'T PANIC.

Graham's chapter

You can read more about Ramsey numbers, including the straightforward proof that six is enough people to guarantee you have three friends or three strangers (that is, $R(3, 3) = 6$), in Imre Leader's lovely article from 2000, 'Friends and strangers', *Plus*. Available at: plus.maths.org/content/friends-and-strangers

INDEX

Entries for individual numbers can be found at the start of the index rather than as words. Page numbers in *italics* denote an illustration

Quercus Editions Ltd
55 Baker Street
7th Floor, South Block
London
W1U 8EW

First published in 2014

Copyright © 2014 Marianne Freiberger and Rachel Thomas

A catalogue record of this book is available from the British Library

ISBN 978 1 78206 154 0

All illustrations by William Donohoe apart from pages 4, 113
(Pikaia Imaging) and 131 (public domain).

Printed and bound in Great Britain by Clays Ltd, St Ives Plc

10 9 8 7 6 5 4 3 2 1